PHYSICS
at a Glance

Tim Mills, BSc
Head of Physics
Brampton College
London

Illustrations by
Cathy Martin

MANSON
PUBLISHING

CONTENTS

ELECTRICAL CIRCUIT SYMBOLS

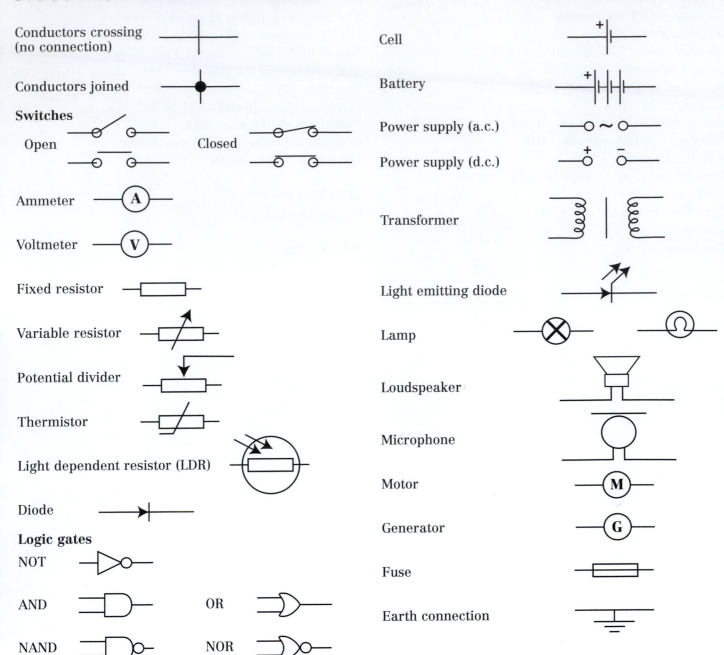

Conductors crossing (no connection)

Conductors joined

Switches

Open

Closed

Ammeter

Voltmeter

Fixed resistor

Variable resistor

Potential divider

Thermistor

Light dependent resistor (LDR)

Diode

Logic gates

NOT

AND

OR

NAND

NOR

Cell

Battery

Power supply (a.c.)

Power supply (d.c.)

Transformer

Light emitting diode

Lamp

Loudspeaker

Microphone

Motor

Generator

Fuse

Earth connection

4

FUNDAMENTAL CONCEPTS

FORCES AND MOTION — Measuring and Describing Motion

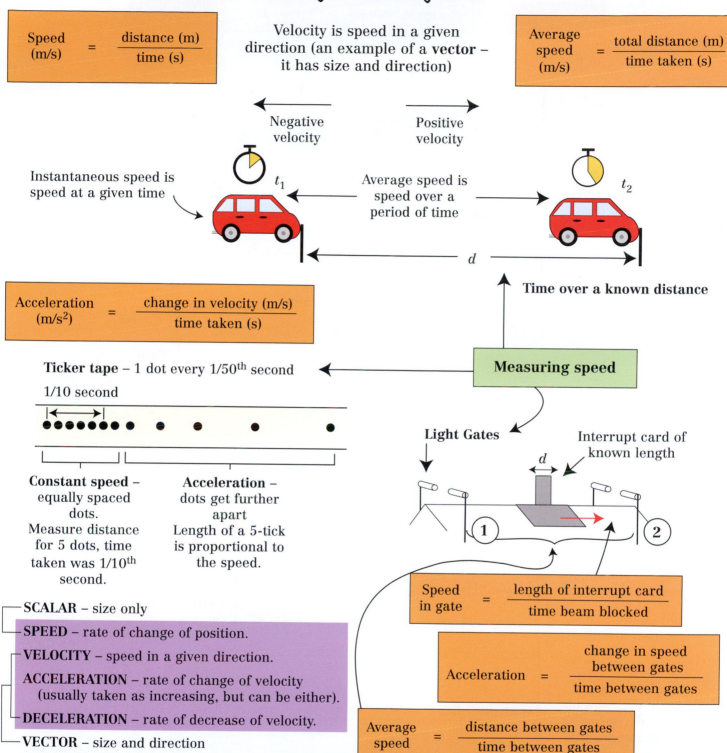

$$\text{Speed (m/s)} = \frac{\text{distance (m)}}{\text{time (s)}}$$

Velocity is speed in a given direction (an example of a **vector** – it has size and direction)

$$\text{Average speed (m/s)} = \frac{\text{total distance (m)}}{\text{time taken (s)}}$$

Negative velocity

Positive velocity

Instantaneous speed is speed at a given time

t_1

Average speed is speed over a period of time

t_2

d

Time over a known distance

$$\text{Acceleration (m/s}^2) = \frac{\text{change in velocity (m/s)}}{\text{time taken (s)}}$$

Measuring speed

Ticker tape – 1 dot every 1/50th second

1/10 second

Constant speed – equally spaced dots. Measure distance for 5 dots, time taken was 1/10th second.

Acceleration – dots get further apart Length of a 5-tick is proportional to the speed.

Light Gates

Interrupt card of known length

d

1   2

$$\text{Speed in gate} = \frac{\text{length of interrupt card}}{\text{time beam blocked}}$$

$$\text{Acceleration} = \frac{\text{change in speed between gates}}{\text{time between gates}}$$

$$\text{Average speed} = \frac{\text{distance between gates}}{\text{time between gates}}$$

- **SCALAR** – size only
- **SPEED** – rate of change of position.
- **VELOCITY** – speed in a given direction.
- **ACCELERATION** – rate of change of velocity (usually taken as increasing, but can be either).
- **DECELERATION** – rate of decrease of velocity.
- **VECTOR** – size and direction

Questions

1. A toy train runs round a circular track of circumference 3 m. After 30 s, it has completed one lap.
   a. What was the train's average speed?
   b. Why is the train's average velocity zero?
   c. The train is placed on a straight track. The train accelerated uniformly from rest to a speed of 0.12 m/s after 10 s. What was its acceleration?
   d. Describe three different ways of measuring the train's average speed and two different ways of measuring the train's instantaneous speed.
   e. How could light gates be used to measure the train's acceleration along a 1 m length of track?
2. Explain the difference between a scalar and vector. Give an example of each.
3. A car leaks oil. One drip hits the road every second. Draw what you would see on the road as the car accelerates.

FORCES AND MOTION Motion Graphs

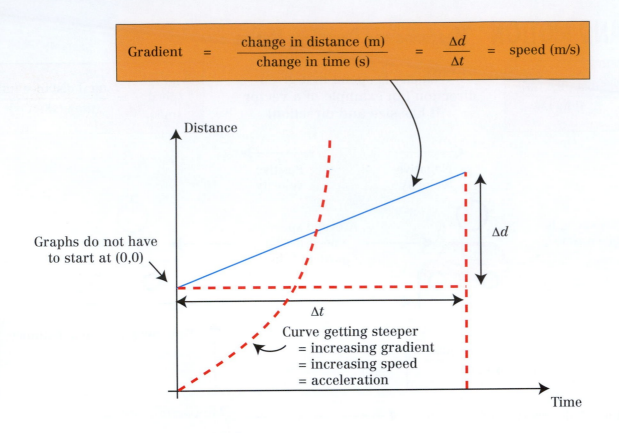

$$\text{Gradient} \quad = \quad \frac{\text{change in distance (m)}}{\text{change in time (s)}} \quad = \quad \frac{\Delta d}{\Delta t} \quad = \quad \text{speed (m/s)}$$

Distance

Graphs do not have to start at (0,0)

Δd

Δt

Curve getting steeper
= increasing gradient
= increasing speed
= acceleration

Time

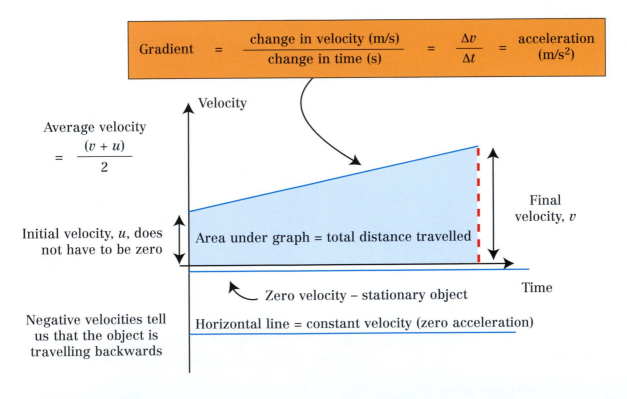

$$\text{Gradient} \quad = \quad \frac{\text{change in velocity (m/s)}}{\text{change in time (s)}} \quad = \quad \frac{\Delta v}{\Delta t} \quad = \quad \text{acceleration (m/s}^2\text{)}$$

Velocity

Average velocity
$= \dfrac{(v + u)}{2}$

Initial velocity, u, does not have to be zero

Area under graph = total distance travelled

Final velocity, v

Time

Zero velocity – stationary object

Negative velocities tell us that the object is travelling backwards

Horizontal line = constant velocity (zero acceleration)

Questions
1. Copy and complete the following sentences:
    a. The slope of a distance – time graph represents _____
    b. The slope of a velocity – time graph represents _____
    c. The area under a velocity – time graph represents _____
2. Redraw the last four graphs from p7 for an object that is decelerating (slowing down).
3. Sketch a distance–time graph for the motion of a tennis ball dropped from a second floor window.
4. Sketch a velocity–time graph for the motion of a tennis ball dropped from a second floor window. Take falling to be a negative velocity and bouncing up to be a positive velocity.

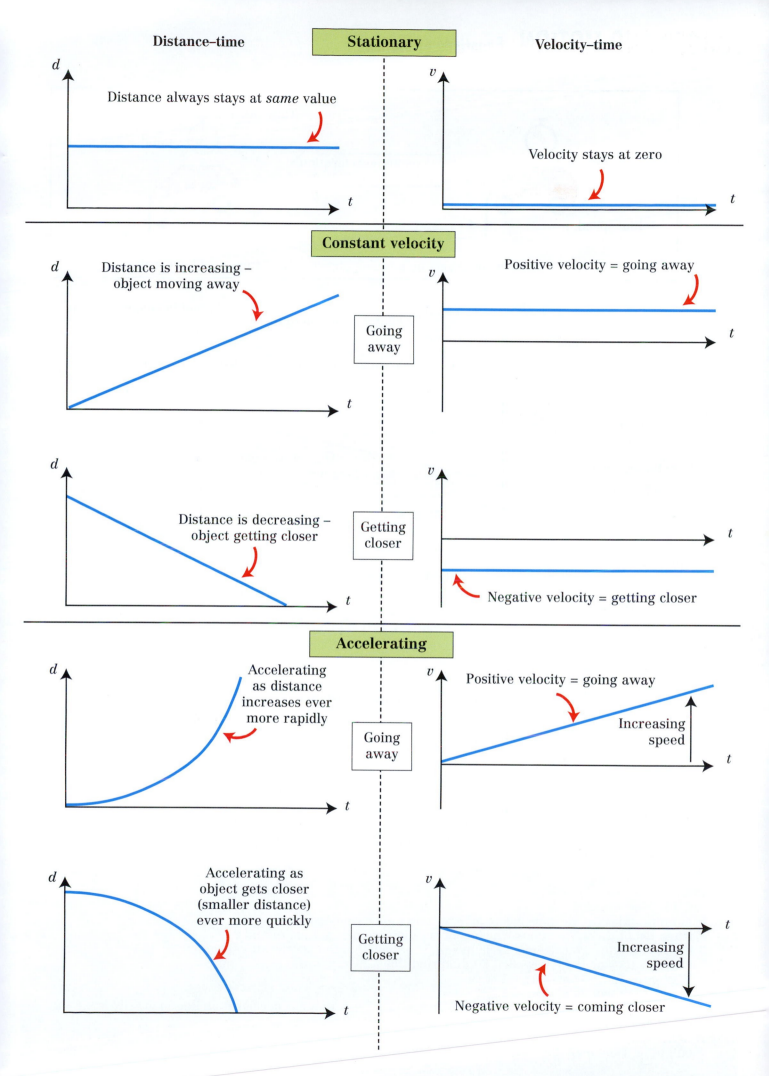

Distance–time

Stationary

Velocity–time

Distance always stays at *same* value

Velocity stays at zero

Constant velocity

Distance is increasing – object moving away

Positive velocity = going away

Going away

Distance is decreasing – object getting closer

Getting closer

Negative velocity = getting closer

Accelerating

Accelerating as distance increases ever more rapidly

Positive velocity = going away

Going away

Increasing speed

Accelerating as object gets closer (smaller distance) ever more quickly

Getting closer

Increasing speed

Negative velocity = coming closer

FORCES AND MOTION Equations of Motion

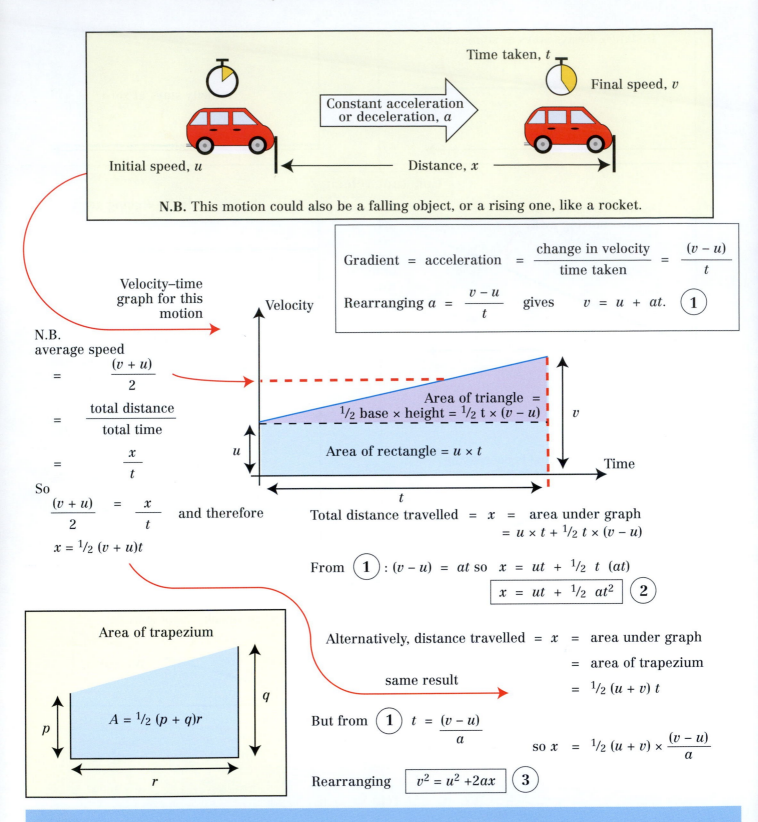

N.B. This motion could also be a falling object, or a rising one, like a rocket.

Velocity–time graph for this motion

Gradient $=$ acceleration $= \dfrac{\text{change in velocity}}{\text{time taken}} = \dfrac{(v - u)}{t}$

Rearranging $a = \dfrac{v - u}{t}$ gives $v = u + at$. (1)

N.B. average speed

$= \dfrac{(v + u)}{2}$

$= \dfrac{\text{total distance}}{\text{total time}}$

$= \dfrac{x}{t}$

So

$\dfrac{(v + u)}{2} = \dfrac{x}{t}$ and therefore

$x = \frac{1}{2}(v + u)t$

Area of triangle $= \frac{1}{2}$ base × height $= \frac{1}{2} t \times (v - u)$

Area of rectangle $= u \times t$

Total distance travelled $= x =$ area under graph
$= u \times t + \frac{1}{2} t \times (v - u)$

From (1): $(v - u) = at$ so $x = ut + \frac{1}{2} t (at)$

$$\boxed{x = ut + \tfrac{1}{2} at^2}\ \ (2)$$

Alternatively, distance travelled $= x =$ area under graph

$=$ area of trapezium

$= \frac{1}{2}(u + v) t$

same result

Area of trapezium

$A = \frac{1}{2}(p + q)r$

But from (1) $t = \dfrac{(v - u)}{a}$

so $x = \frac{1}{2}(u + v) \times \dfrac{(v - u)}{a}$

Rearranging $\boxed{v^2 = u^2 + 2ax}\ \ (3)$

Questions
Show ALL your working.
1. What quantities do the variables x, u, v, a, and t each represent?
2. Write a list of three equations which connect the variables x, u, v, a, and t.
3. A car accelerates from 10 m/s to 22 m/s in 5 s. Show that the acceleration is about 2.5 m/s^2.
4. Now show the car in (3) travelled 80 m during this acceleration:
   a. Using the formula $v^2 = u^2 + 2ax$.
   b. Using the formula $x = ut + \frac{1}{2}at^2$.
5. A ball falls from rest. After 4 s, it has fallen 78.4 m. Show that the acceleration due to gravity is 9.8 m/s^2.
6. Show that $x = \frac{1}{2}(u + v)(v - u)/a$ rearranges to $v^2 = u^2 + 2ax$.
7. A ball thrown straight up at 15 m/s, feels a downward acceleration of 9.8 m/s^2 due to the pull of the Earth on it. How high does the ball go before it starts to fall back?

FORCES AND MOTION   Describing Forces

Push [box]

FORCES
- Description
- Are vectors

• Type of force → Gravitational pull
• Caused by → of the Earth
• Acting on → on the moon

E.g.

[box] **Pull**

Rotation   F   F   **Change shape**   F

EFFECTS OF FORCES

Speed up ← [box] ← F

Slow down [box] ← F

Twist

Direction   F   F   F
Shown by direction of arrow

Size
Shown by size/length of arrow

Friction   Gravitational
Electrostatic   **TYPES OF FORCE**   Tension
Normal contact   Magnetic   Thrust   Drag

N.B. not the forces caused by the object acting on another object

Free body diagrams
E.g. Sliding box

Simple diagrams to show all the forces acting *on* an object.

Push of floor *on* box   properly called → **Normal contact force**

Push of floor = weight (arrows same length but act in opposite directions)

Friction of floor *on* box

Weight (gravitational pull of Earth *on* box)

Push of one surface on another at right angles to the surface.
Due to atoms in each surface being slightly squashed together and pushing back

Weight – shorthand for 'the gravitational pull of a planet or moon (usually the Earth) on an object'

E.g. pendulum   Tension (pull of string on bob)

Weight (gravitational pull of Earth on bob)

> **Resultant force** – a single force that can replace all the forces acting on a body and have the same overall effect as all the individual forces acting together.
> It is the sum of all the individual forces taking their directions into account.

E.g. rocket

Acceleration (double-headed arrow not attached to object)

Thrust (push of gas on rocket)

Velocity (arrow not attached to object)

Weight (gravitational pull of Earth on rocket)

Parallel forces – add

Push of wind on boat
Drag of water on boat

Resultant = push of wind on boat – drag of water on boat

Shows opposite direction

Perpendicular forces – Pythagoras Theorem

Tension 1   E.g. Bow and arrow

Tension 2   Resultant force on arrow, R

$R^2 = (\text{tension 1})^2 + (\text{tension 2})^2$

Questions
1. List three effects forces can have.
2. Explain what the term 'resultant force' means.
3. To describe a force fully, what three pieces of information should be recorded?
4. Copy and add arrows to these diagrams to show all the forces (and their directions) acting on:
   a. A netball flying through the air.
   b. A jet ski.
   c. A cyclist freewheeling down a hill.
   d. A child on a swing pushed by an adult.
5. Calculate the resultant force in the following cases:

   a.   3 N ← [box] → 5 N

   b.   4 N ↑   7 N ← [box] → 10 N   4 N ↓

   c.   3 N ↑   [box] → 4 N

   d.   4 N ↑ 8 N ↑   [box] → 5 N

   e.   1 N ↑ 3 N ↑ 2 N   10 N ← [box] → 4 N   7 N ↓

9

FORCES AND MOTION Balanced Forces – Newton's First Law

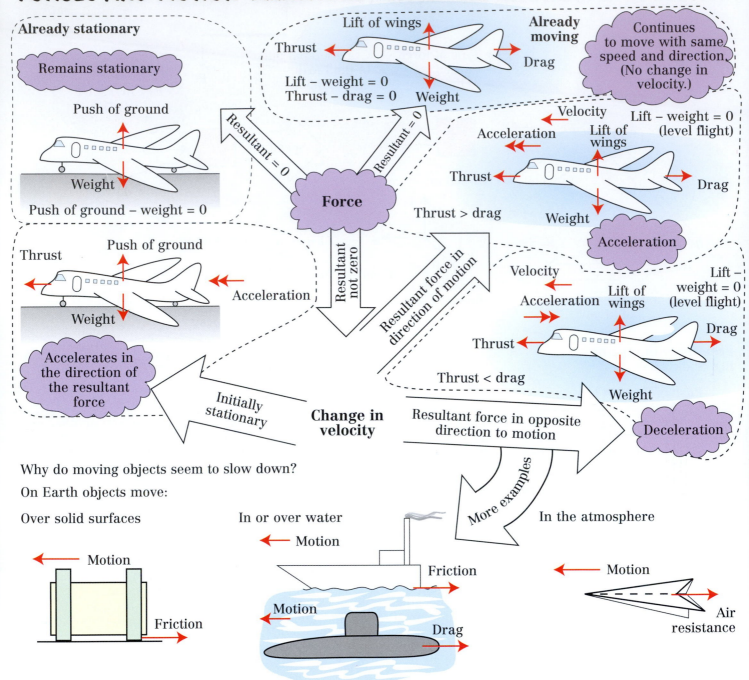

Already stationary

Remains stationary

Push of ground

Weight

Push of ground – weight = 0

Thrust

Push of ground

Acceleration

Weight

Accelerates in the direction of the resultant force

Lift of wings

Thrust

Drag

Lift – weight = 0
Thrust – drag = 0

Weight

Already moving

Continues to move with same speed and direction. (No change in velocity.)

Force

Resultant = 0

Resultant = 0

Resultant not zero

Resultant force in direction of motion

Initially stationary

Change in velocity

Velocity

Acceleration

Lift of wings

Thrust

Weight

Drag

Lift – weight = 0 (level flight)

Thrust > drag

Acceleration

Velocity

Acceleration

Lift of wings

Thrust

Weight

Drag

Lift – weight = 0 (level flight)

Thrust < drag

Resultant force in opposite direction to motion

Deceleration

Why do moving objects seem to slow down?

On Earth objects move:

Over solid surfaces

Motion

Friction

In or over water

Motion

Friction

Motion

Drag

More examples

In the atmosphere

Motion

Air resistance

In all cases, resistive forces act to oppose motion. Therefore, unless a force is applied to balance the resistive force the object will slow down. In space, there are no resistive forces and objects will move at constant speed in a straight line unless another force acts.

Newton's First Law of Motion:
- If the resultant force acting on a body is zero, it will remain at rest or continue to move at the same speed in the same direction.
- If the resultant force acting on a body is not zero, it will accelerate in the direction of the resultant force.

Questions
1. In which of the following situations is the resultant force zero? Explain how you decided.
   a. A snooker ball resting on a snooker table.
   b. A car accelerating away from traffic lights.
   c. A ball rolling along level ground and slowing down.
   d. A skier travelling down a piste at constant speed.
   e. A toy train travelling round a circular track at constant speed.
2. A lift and its passengers have a weight of 5000 N. Is the tension in the cable supporting the lift:

i. Greater than 5000 N, ii. Less than 5000 N, iii. Exactly 5000 N when:
   a. The lift is stationary?
   b. Accelerating upwards?
   c. Travelling upwards at a constant speed?
   d. Decelerating whilst still travelling upwards?
   e. Accelerating downwards?
   f. Travelling downwards at constant velocity?
   g. Decelerating while still travelling downward?
3. Explain why all objects moving on Earth will eventually come to rest unless another force is applied?

FORCES AND MOTION Unbalanced Forces – Newton's Second Law

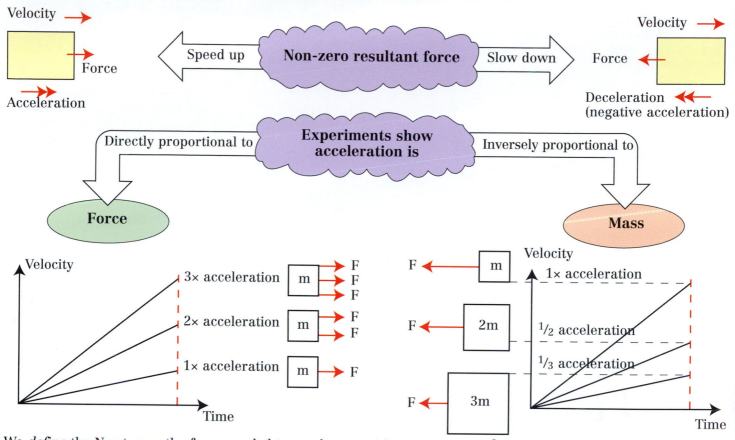

We *define* the Newton as the force needed to accelerate a 1 kg mass at 1 m/s². Therefore, we can write:

Newton's Second Law

Force (N) = mass (kg) × acceleration (m/s²).

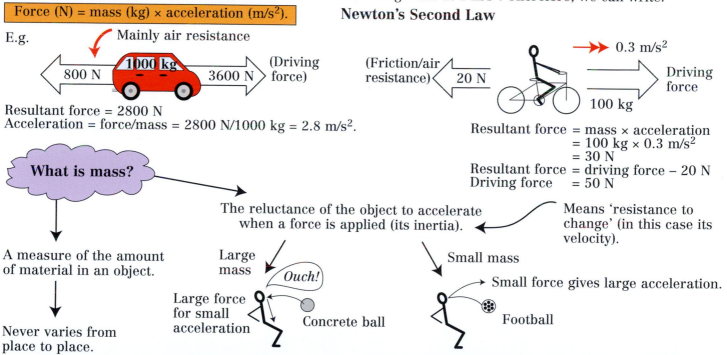

E.g.

1000 kg

Mainly air resistance

800 N   3600 N (Driving force)

Resultant force = 2800 N
Acceleration = force/mass = 2800 N/1000 kg = 2.8 m/s².

0.3 m/s²

(Friction/air resistance) 20 N   Driving force

100 kg

Resultant force = mass × acceleration
= 100 kg × 0.3 m/s²
= 30 N
Resultant force = driving force – 20 N
Driving force = 50 N

What is mass?

The reluctance of the object to accelerate when a force is applied (its inertia).

Means 'resistance to change' (in this case its velocity).

A measure of the amount of material in an object.

Large mass

Large force for small acceleration

Ouch!

Concrete ball

Small mass

Small force gives large acceleration.

Football

Never varies from place to place.

Questions
1. Calculate:
   a. The force needed to accelerate a 70 kg sprinter at 6 m/s².
   b. The acceleration of a 10 g bullet with 2060 N explosive force in a gun barrel.
   c. The mass of a ship accelerating at 0.09 m/s² with a resultant thrust of 6 400 000 N from the propellers.
2. An underground tube train has mass of 160 000 kg and can produce a maximum driving force of 912 000 N.
   a. When accelerating in the tunnel using the maximum driving force show the acceleration should be 5.7 m/s².
   b. In reality, the acceleration is only 4.2 m/s². Hence show the resistive forces on the train are 240 000 N.
3. Explain why towing a caravan reduces the maximum acceleration of a car (two reasons).
4. A football made of concrete would be weightless in deep space. However, it would not be a good idea for an astronaut to head it. Why not?

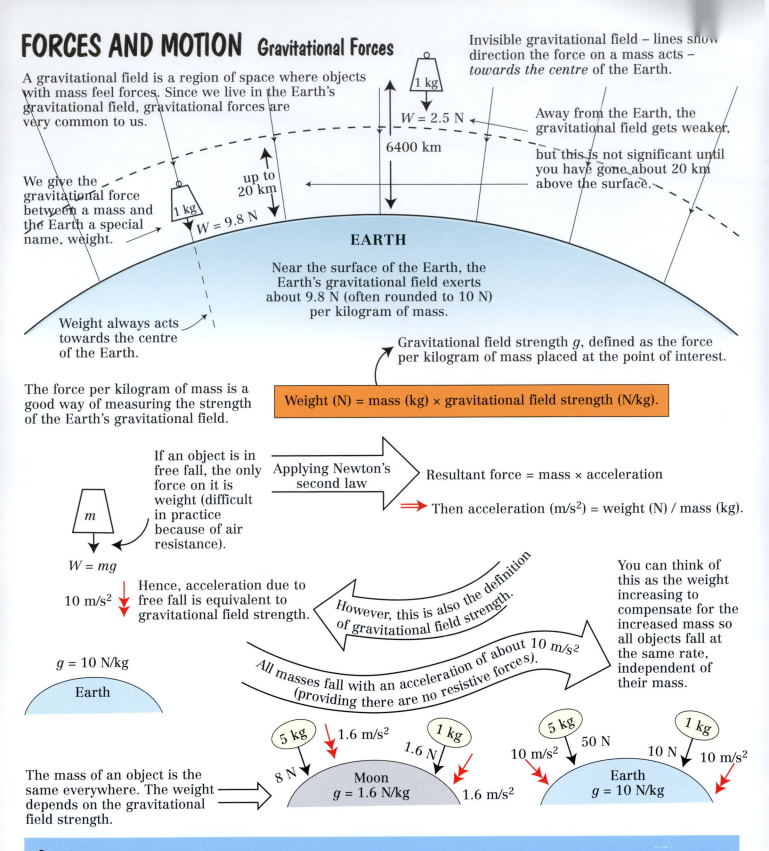

A gravitational field is a region of space where objects with mass feel forces. Since we live in the Earth's gravitational field, gravitational forces are very common to us.

Invisible gravitational field – lines show direction the force on a mass acts – *towards the centre* of the Earth.

1 kg

W = 2.5 N

6400 km

Away from the Earth, the gravitational field gets weaker,

but this is not significant until you have gone about 20 km above the surface.

We give the gravitational force between a mass and the Earth a special name, weight.

up to 20 km

1 kg

W = 9.8 N

EARTH

Weight always acts towards the centre of the Earth.

Near the surface of the Earth, the Earth's gravitational field exerts about 9.8 N (often rounded to 10 N) per kilogram of mass.

Gravitational field strength g, defined as the force per kilogram of mass placed at the point of interest.

The force per kilogram of mass is a good way of measuring the strength of the Earth's gravitational field.

Weight (N) = mass (kg) × gravitational field strength (N/kg).

If an object is in free fall, the only force on it is weight (difficult in practice because of air resistance).

m

Applying Newton's second law

Resultant force = mass × acceleration

Then acceleration (m/s^2) = weight (N) / mass (kg).

$W = mg$

10 m/s^2

Hence, acceleration due to free fall is equivalent to gravitational field strength.

However, this is also the definition of gravitational field strength.

All masses fall with an acceleration of about 10 m/s^2 (providing there are no resistive forces).

You can think of this as the weight increasing to compensate for the increased mass so all objects fall at the same rate, independent of their mass.

$g = 10$ N/kg

Earth

5 kg — 1.6 m/s^2

8 N

1 kg

1.6 N

5 kg — 50 N

10 m/s^2

1 kg

10 N

10 m/s^2

The mass of an object is the same everywhere. The weight depends on the gravitational field strength.

Moon
$g = 1.6$ N/kg

1.6 m/s^2

Earth
$g = 10$ N/kg

Questions
1. Near the surface of the Earth, what are the values of:
   a. The acceleration due to free fall?
   b. The gravitational field strength?
2. What are the weights on the Earth of:
   a. A book of mass 2 kg?
   b. An apple of mass 100 g?
   c. A girl of mass 60 kg?
   d. A blade of grass of mass 0.1 g?
3. What would the masses and weights of the above objects be on the moon? (Gravitational field strength on the moon = 1.6 N/kg).
4. 6400 km above the surface of the Earth a 1 kg mass has a weight of 2.5 N. What is the gravitational field strength here? If the mass was dropped, and started falling towards the centre of the Earth, what would its initial acceleration be?
5. Write a few sentences to explain the difference between mass and weight.

FORCES AND MOTION

Terminal Velocity

Terminal velocity occurs when the accelerating and resistive force on an object are balanced.

Key ideas:
- Drag/resistive forces on objects increase with increasing speed for objects moving through a fluid, e.g. air or water.
- When accelerating and resistive forces are balanced, Newton's First Law says that the object will continue to travel at constant velocity.

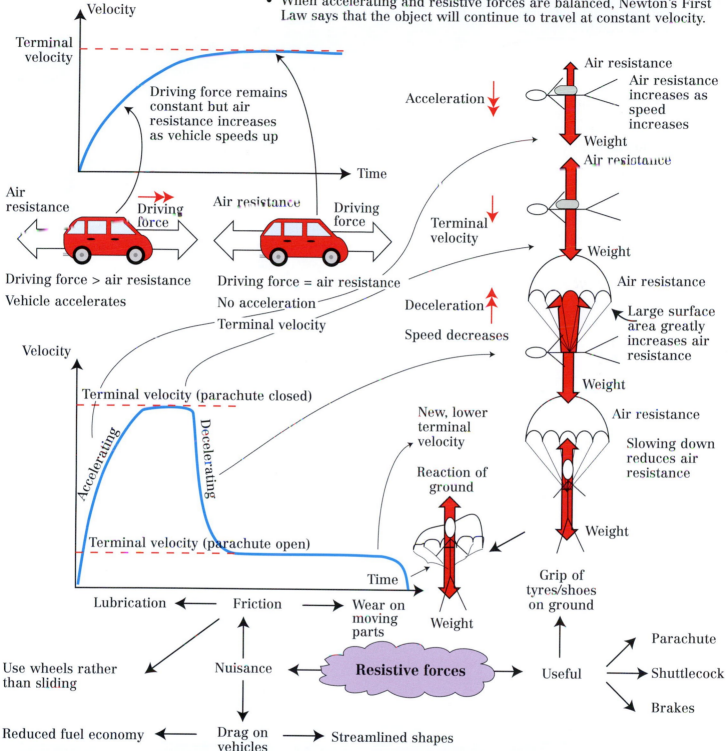

Driving force remains constant but air resistance increases as vehicle speeds up

Air resistance — Driving force

Air resistance — Driving force

Driving force > air resistance
Vehicle accelerates

Driving force = air resistance
No acceleration
Terminal velocity

Velocity

Terminal velocity (parachute closed)

Accelerating

Decelerating

Terminal velocity (parachute open)

Time

Acceleration

Air resistance
Air resistance increases as speed increases
Weight

Terminal velocity

Air resistance
Weight

Deceleration
Speed decreases

Air resistance
Large surface area greatly increases air resistance
Weight

New, lower terminal velocity

Air resistance
Slowing down reduces air resistance
Weight

Reaction of ground

Weight

Lubrication ← Friction → Wear on moving parts

Use wheels rather than sliding

Nuisance

Resistive forces

Useful

Grip of tyres/shoes on ground

Parachute

Shuttlecock

Brakes

Reduced fuel economy ← Drag on vehicles → Streamlined shapes

Questions

1. What happens to the size of the drag force experienced by an object moving through a fluid (e.g. air or water) as it speeds up?
2. What force attracts all objects towards the centre of the Earth?
3. Why does a car need to keep its engine running to travel at constant velocity?
4. A hot air balloon of weight 6000 N is released from its mooring ropes.
   a. The upward force from the hot air rising is 6330 N. Show the initial acceleration is about 0.5 m/s^2
   b. This acceleration gradually decreases as the balloon rises until it is travelling at a constant velocity. Explain why.
   c. A mass of 100 kg is thrown overboard. What will happen to the balloon now?
   d. Sketch a velocity–time graph for the whole journey of the balloon as described in parts a–c.
5. Explain why the following are likely to increase the petrol consumption of a car:
   a. Towing a caravan.
   b. Adding a roof rack.
   c. Driving very fast.

FORCES AND MOTION

Projectiles

The secret is to consider the velocity of the projectile to be made up of horizontal and vertical velocities, which can be considered separately.

A true projectile only has one force, weight, acting on it when it is in flight.

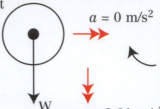

$a = 0$ m/s^2

We ignore air resistance; therefore, there is no horizontal acceleration (or deceleration).

W

$g = -9.81$ m/s^2

At any time motion is made up of

1. Horizontal velocity: No horizontal forces (ignoring air resistance). Therefore by Newton's First Law, no change in velocity horizontally.

2. Vertical velocity: Projectile accelerates downwards under gravity, slowing as it rises, stopping at the top and falling back.

Use $v_V = u_V + gt$ and $x_V = u_V t + \frac{1}{2} gt^2$ where $g = -9.81$ m/s^2

Examples:

Kicked football

Golf ball

Darts

Cannon ball

Long jumper

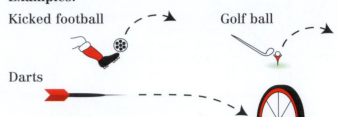

NOT rockets

Thrust

Two forces

Weight

Initial velocity.

u_V

v_V

v_H

v_H

$v_{resultant}$

x_V

$a = 0$ m/s^2

$g = -9.81$ m/s^2

v_V

v_H

v_H

Range = V_H × time of flight

Time of flight = 2× time to maximum height.

Path is called the *trajectory*. Shape is *parabolic*, same as the graph of $y = c - x^2$

v_H

v_V

y

c

x

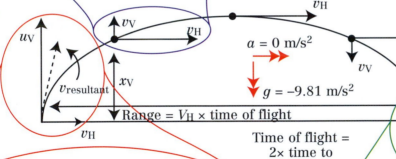

Initial velocity is made up from two vectors at right angles, called components.
The overall effect (the *resultant*) is the initial velocity of the projectile and is found by Pythagoras' Theorem.

u_V

v_R

v_H

$v_R = \sqrt{u_V + v_H}$

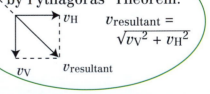

Impact velocity by Pythagoras' Theorem.

v_H

v_V

$v_{resultant}$

$v_{resultant} = \sqrt{v_V^2 + v_H^2}$

Questions
1. In an ideal world how many forces act on a projectile, and what are they?
2. State the value of the vertical acceleration of a projectile.
3. Explain why the horizontal acceleration of a projectile is zero. What assumption has to be made?
4. Explain why a firework rocket cannot be analysed as a projectile with the methods shown here.
5. A ball is kicked so it has a velocity of 15.59 m/s horizontally and 9.0 m/s vertically.

a. Show that the resultant velocity of the ball has a magnitude of 18.0 m/s.
b. Show that the ball takes 0.92 s to reach its maximum height above the ground.
c. For how long in total is the ball in the air and how far along the ground will it travel?
d. Show the maximum height the ball reaches is 4.1 m.
e. What will the magnitude of its resultant velocity be when it hits the ground? Hint: no calculation needed.

FORCES AND MOTION Newton's Third Law

Whenever an object experiences a force it always exerts an equal . . . . . and . . . . . opposite force on the object causing the force.

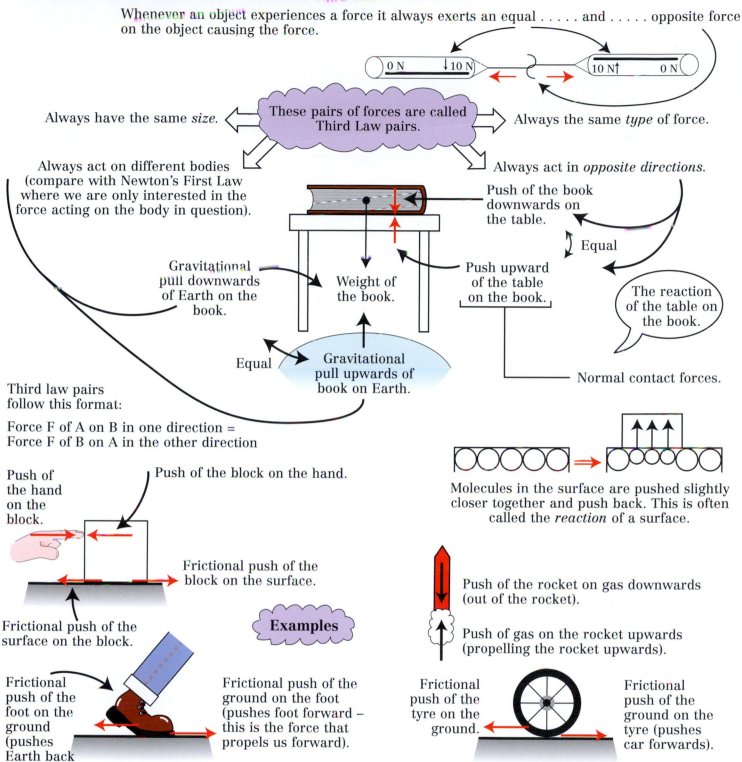

0 N ↓10 N   10 N↑ 0 N

These pairs of forces are called Third Law pairs.

Always have the same *size*.

Always the same *type* of force.

Always act on different bodies (compare with Newton's First Law where we are only interested in the force acting on the body in question).

Always act in *opposite directions*.

Push of the book downwards on the table.

Gravitational pull downwards of Earth on the book.

Weight of the book.

Push upward of the table on the book.

Equal

The reaction of the table on the book.

Equal

Gravitational pull upwards of book on Earth.

Normal contact forces.

Third law pairs follow this format:

Force F of A on B in one direction = Force F of B on A in the other direction

Push of the block on the hand.

Push of the hand on the block.

Frictional push of the block on the surface.

Frictional push of the surface on the block.

Molecules in the surface are pushed slightly closer together and push back. This is often called the *reaction* of a surface.

Examples

Frictional push of the foot on the ground (pushes Earth back slightly).

Frictional push of the ground on the foot (pushes foot forward – this is the force that propels us forward).

Push of the rocket on gas downwards (out of the rocket).

Push of gas on the rocket upwards (propelling the rocket upwards).

Frictional push of the tyre on the ground.

Frictional push of the ground on the tyre (pushes car forwards).

If the ground is icy, both these forces are very small and we cannot walk or drive forwards.

Questions

1. Explain what is meant by the term 'normal contact force'.
2. A jet engine in an aircraft exerts 200 000 N on the exhaust gases. What force do the gases exert on the aircraft?
3. Describe the force that forms a Third Law pair with the following. In each case, draw a diagram to illustrate the two forces:
   a. The push east of the wind on a sail.
   b. The push left of a bowstring on an arrow.
   c. The frictional push south of a train wheel on a rail.
   d. The normal contact force downwards of a plate on a table.
   e. The attraction right of the north magnetic pole of a bar magnet on a south magnetic pole of a different magnet.
4. Why are the following not Third Law pairs? (There may be more than one reason for each.)
   a. The weight of a mug sitting on a table; the normal contact force of the tabletop on the mug.
   b. The weight of the passengers in a lift car; the upward tension in the lift cable.
   c. The weight of a pool ball on a table; the horizontal push of the cue on the ball.
   d. The attraction between the north and south magnetic poles of the same bar magnet.
5. Explain why it is very difficult (and dangerous) to ride a bicycle across a sheet of ice.

FORCES AND MOTION Momentum and Force (Newton's Laws revisited)

Momentum helps to describe how moving objects will behave.

Momentum (kgm/s) = mass (kg) × velocity (m/s)

Momentum is a vector. It has size and direction (the direction of the velocity).

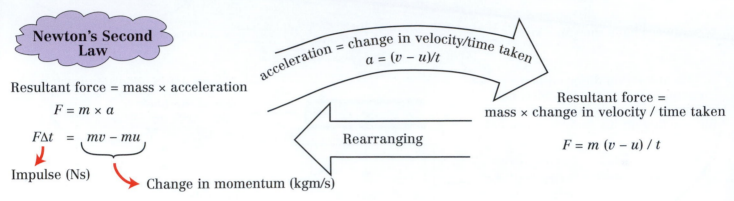

Newton's Second Law

Resultant force = mass × acceleration

$$F = m \times a$$

acceleration = change in velocity/time taken
$$a = (v - u)/t$$

Resultant force = mass × change in velocity / time taken

$$F = m(v - u)/t$$

Rearranging

$$F\Delta t = \underbrace{mv - mu}$$

Impulse (Ns)

Change in momentum (kgm/s)

Hence, an alternative version of Newton's Second Law

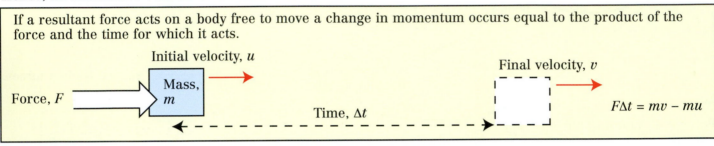

If a resultant force acts on a body free to move a change in momentum occurs equal to the product of the force and the time for which it acts.

Initial velocity, u

Force, F

Mass, m

Time, Δt

Final velocity, v

$$F\Delta t = mv - mu$$

Also consider

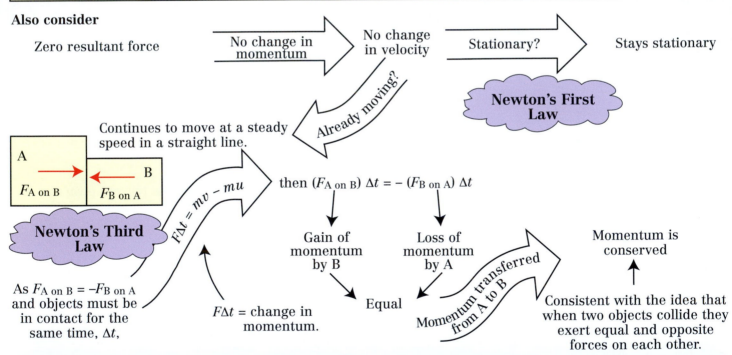

Zero resultant force → No change in momentum → No change in velocity → Stationary? → Stays stationary

Already moving?

Newton's First Law

A
$F_{A \text{ on } B}$
B
$F_{B \text{ on } A}$

Continues to move at a steady speed in a straight line.

$$F\Delta t = mv - mu$$

then $(F_{A \text{ on } B}) \Delta t = - (F_{B \text{ on } A}) \Delta t$

Newton's Third Law

As $F_{A \text{ on } B} = -F_{B \text{ on } A}$ and objects must be in contact for the same time, Δt,

$$F\Delta t = \text{change in momentum.}$$

Gain of momentum by B

Loss of momentum by A

Equal

Momentum transferred from A to B

Momentum is conserved

Consistent with the idea that when two objects collide they exert equal and opposite forces on each other.

Questions

1. What units do we use to measure momentum and impulse (2 answers)?
2. Calculate the momentum of:
   a. A 55 kg girl running at 7 m/s north.
   b. A 20 000 kg aircraft flying at 150 m/s south.
   c. A 20 g snail moving at 0.01 m/s east.
3. What is the connection between force and change in momentum?

4. What is the change in momentum in the following cases:
   a. A 5 N force acting for 10 s?
   b. A 500 N force acting for 0.01 s?
5. What force is required to:
   a. Accelerate a 70 kg athlete from 0 to 9 m/s in 2 s?
   b. Accelerate a 1000 kg car from rest to 26.7 m/s in 5 s?
   c. Stop a 10 g bullet travelling at 400 m/s in 0.001 s?

6. What would be the effect on the force needed to change momentum if the time the force acts for is increased?
7. A 2564 kg space probe is to be accelerated from 7.7 km/s to 11.0 km/s. If it has a rocket motor that can produce 400 N of thrust, for how long would it need to burn assuming that no resistive forces act? Why might this not be practical? How else might the space probe gain sufficient momentum (see p113 for ideas)?

FORCES AND MOTION Momentum Conservation and Collisions

Law of Conservation of Momentum:

> Momentum cannot be created or destroyed but can be transferred from one object to another when they interact.

There are no exceptions to this. It is applied to analyse interactions between objects, which can be classified as:

Velocities

Up and **right** are taken as **positive**.

Down and **left** are taken as **negative**.

Or

Objects initially moving towards each other

Objects originally stationary and move apart

Collisions

or

Explosions

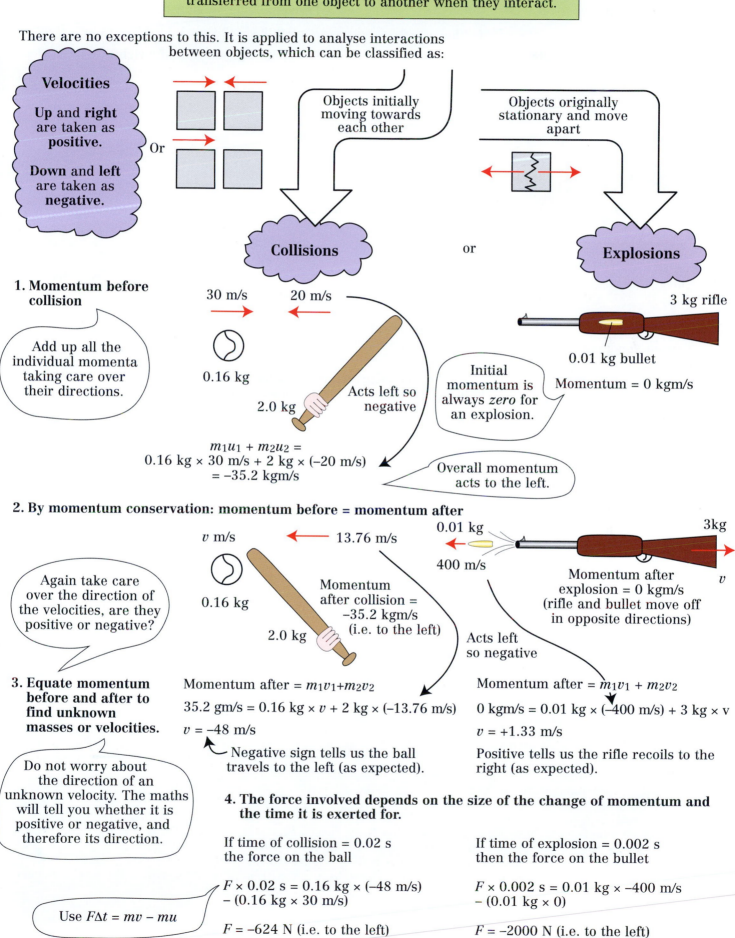

1. Momentum before collision

30 m/s   20 m/s

Add up all the individual momenta taking care over their directions.

0.16 kg

2.0 kg   Acts left so negative

3 kg rifle

0.01 kg bullet

Momentum = 0 kgm/s

Initial momentum is always *zero* for an explosion.

$m_1u_1 + m_2u_2 =$
0.16 kg × 30 m/s + 2 kg × (−20 m/s)
= −35.2 kgm/s

Overall momentum acts to the left.

2. By momentum conservation: momentum before = momentum after

v m/s   ← 13.76 m/s

0.01 kg   3kg

400 m/s

Again take care over the direction of the velocities, are they positive or negative?

0.16 kg

2.0 kg

Momentum after collision = −35.2 kgm/s (i.e. to the left)

Momentum after explosion = 0 kgm/s (rifle and bullet move off in opposite directions)

v

Acts left so negative

3. Equate momentum before and after to find unknown masses or velocities.

Momentum after = $m_1v_1 + m_2v_2$

35.2 gm/s = 0.16 kg × v + 2 kg × (−13.76 m/s)

$v = −48$ m/s

Negative sign tells us the ball travels to the left (as expected).

Momentum after = $m_1v_1 + m_2v_2$

0 kgm/s = 0.01 kg × (−400 m/s) + 3 kg × v

$v = +1.33$ m/s

Positive tells us the rifle recoils to the right (as expected).

Do not worry about the direction of an unknown velocity. The maths will tell you whether it is positive or negative, and therefore its direction.

4. The force involved depends on the size of the change of momentum and the time it is exerted for.

If time of collision = 0.02 s the force on the ball

If time of explosion = 0.002 s then the force on the bullet

Use $F\Delta t = mv - mu$

$F × 0.02$ s = 0.16 kg × (−48 m/s) − (0.16 kg × 30 m/s)

$F = −624$ N (i.e. to the left)

$F × 0.002$ s = 0.01 kg × −400 m/s − (0.01 kg × 0)

$F = −2000$ N (i.e. to the left)

FORCES AND MOTION Momentum Conservation and Collisions (continued)

The calculation of the force exerted on the bullet and the ball would work equally well if the force on the bat or the rifle were calculated. The size of the force would be the same, but in the opposite direction according to Newton's Third Law. Again using $F\Delta t = mv - mu$.

Force of ball on bat

$F \times 0.02$ s $= 2$ kg $\times (-13.76$ m/s$) - 2$ kg $(-20$m/s$)$

$F = 624$ N (positive, to the right).

Force of bullet on gun

$F \times 0.002$ s $= (3$ kg $\times 1.33$ m/s$) - (3$ kg $\times 0$ m/s$)$

$F = 2000$ N (positive, to the right).

These calculations show that the force involved depends on.

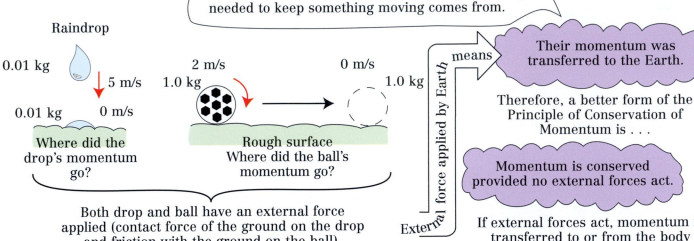

Both 1 kg

Metal head   *vs.*   Wooden head

Short impact time – larger force.

Long impact time – less force.

Time of impact

Change of momentum

Sledge hammer = 10 kg

Both metal heads have the same contact time.

Light hammer = 1 kg

Larger change of momentum exerts a larger force.

Sometimes it appears that momentum is not conserved.

This is where the incorrect idea of a force being needed to keep something moving comes from.

Raindrop

0.01 kg   2 m/s   0 m/s
       ↓5 m/s   1.0 kg   1.0 kg

0.01 kg   0 m/s

Where did the drop's momentum go?

Rough surface
Where did the ball's momentum go?

Both drop and ball have an external force applied (contact force of the ground on the drop and friction with the ground on the ball).

External force applied by Earth

means

Their momentum was transferred to the Earth.

Therefore, a better form of the Principle of Conservation of Momentum is . . .

Momentum is conserved provided no external forces act.

If external forces act, momentum is transferred to or from the body exerting the force.

Questions
1. When a raindrop hits the ground where does its momentum go?
2. Why do boxers wear padded gloves?
3. A squash ball is hit against a wall and bounces off. An equal mass of plasticine is thrown at the same wall with the same speed as the ball, but it sticks on impact. Which exerts the larger force on the wall and why?
4. A golfer swings a 0.2 kg club at 45 m/s. It hits a stationary golf ball of mass 45 g, which leaves the tee at 65 m/s.
   a. What was the momentum of the club before the collision?
   b. What was the momentum of the ball after the collision?
   c. Hence, show that the club's velocity is about 30 m/s after the collision.
   d. If the club is in contact with the ball for 0.001 s, what is the average force the club exerts on the ball?
5. A 1.5 kg air rifle fires a 1 g pellet at 150 m/s. What is the recoil velocity of the rifle? Show that the force exerted by the rifle on the pellet is about 70 N if the time for the pellet to be fired is 0.0021 s.
6. Assume that the average mass of a human being is 50 kg. If all 5.5×10^9 humans on Earth stood shoulder to shoulder in one place, and jumped upward at 1 m/s with what velocity would the Earth, mass 6×10^{24} kg recoil?
7. Two friends are ice-skating. One friend with mass 70 kg is travelling at 4 m/s. The other of mass 60 kg travelling at 6 m/s skates up behind the first and grabs hold of them. With what speed will the two friends continue to move while holding onto each other?

FORCES AND MOTION Motion in Circles and Centripetal Forces

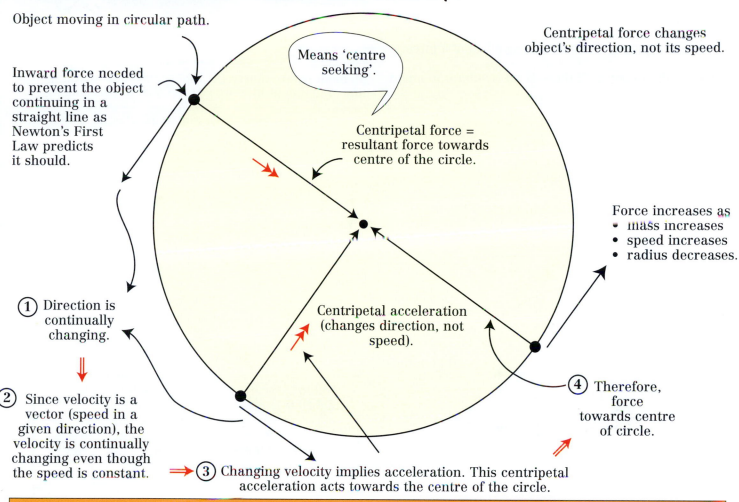

Object moving in circular path.

Inward force needed to prevent the object continuing in a straight line as Newton's First Law predicts it should.

Means 'centre seeking'.

Centripetal force changes object's direction, not its speed.

Centripetal force = resultant force towards centre of the circle.

Force increases as
- mass increases
- speed increases
- radius decreases.

① Direction is continually changing.

② Since velocity is a vector (speed in a given direction), the velocity is continually changing even though the speed is constant.

Centripetal acceleration (changes direction, not speed).

④ Therefore, force towards centre of circle.

⇒ ③ Changing velocity implies acceleration. This centripetal acceleration acts towards the centre of the circle.

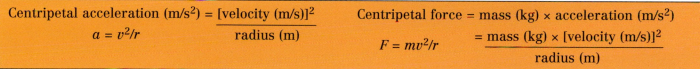

Centripetal acceleration (m/s^2) = $\dfrac{[\text{velocity (m/s)}]^2}{\text{radius (m)}}$

$a = v^2/r$

Centripetal force = mass (kg) × acceleration (m/s^2)

$= \dfrac{\text{mass (kg)} \times [\text{velocity (m/s)}]^2}{\text{radius (m)}}$

$F = mv^2/r$

Centripetal force is not a force in its own right – it must be *provided* by another type of force.

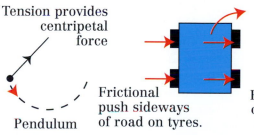

Tension provides centripetal force

Pendulum

Frictional push sideways of road on tyres.

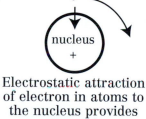

Electrostatic attraction of electron in atoms to the nucleus provides centripetal force.

e^-

nucleus +

Normal contact force on clothes in washing machine drum provides centripetal force.

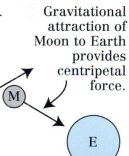

Gravitational attraction of Moon to Earth provides centripetal force.

M

E

Questions
1. What force provides the centripetal force in each of these cases?
   a. The Earth moving in orbit around the Sun.
   b. Running around a sharp bend.
   c. A child on a swing.
2. Explain how a passenger on a roundabout at a funfair can be moving at constant speed around the circle and yet accelerating. In what direction is the acceleration?
3. What is the centripetal acceleration of, and force on, the following:
   a. A wet sweater of mass 1 kg, spinning in a washing machine drum of radius 35 cm, moving at 30 m/s.
   b. A snowboarder of mass 70 kg travelling round a half pipe of radius 6 m at 5 m/s.

4. The Earth has a mass of 6×10^{24} kg. Its orbit radius is 1.5×10^{11} m and the gravitational attraction to the Sun is 3.6×10^{22} N.
   a. Show that the circumference of the Earth's orbit is about 9.5×10^{11} m.
   b. Show that the Earth's speed around the Sun is about 30 000 m/s.
   c. Therefore, show that the time to orbit the Sun is about 3×10^7 s.
   d. Show that this is about 365 days.
5. On a very fast rotating ride at a funfair, your friend says that they feel a force trying to throw them sideways out of the ride. How would you convince your friend that actually they are experiencing a force pushing *inwards*? You should refer to Newton's First and Third Laws in your explanation.

FORCES AND MOTION
Moments and Stability

A moment (or torque) is the turning effect of a force.

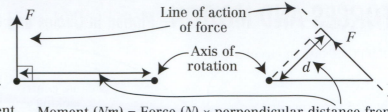

A body will not rotate if there is no resultant moment.

Moment (*Nm*) = Force (*N*) × perpendicular distance from line of action of the force to the axis of rotation (*m*).

Anticlockwise moment $F \times d$ = Clockwise moment $f \times D$

$(4\ N \times 2.4\ m) = (3\ N \times 1.2\ m)$ $=$ $(6\ N \times 1\ m) + (2\ N \times 3.6\ m)$
$13.2\ Nm$ $=$ $13.2\ Nm$

Centre of mass:

You could think of the mass behaving as if it were all concentrated here.

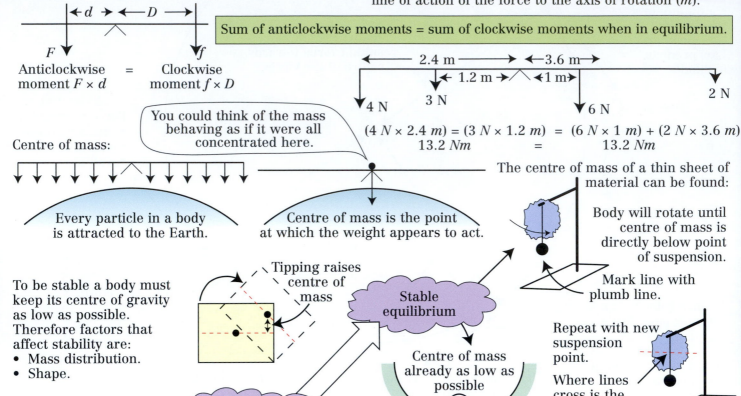

Every particle in a body is attracted to the Earth.

Centre of mass is the point at which the weight appears to act.

The centre of mass of a thin sheet of material can be found:

Body will rotate until centre of mass is directly below point of suspension.

Mark line with plumb line.

To be stable a body must keep its centre of gravity as low as possible. Therefore factors that affect stability are:
• Mass distribution.
• Shape.

Tipping raises centre of mass

Stable equilibrium

Centre of mass already as low as possible

Repeat with new suspension point.

Where lines cross is the centre of mass as it is the only point that is on all the lines.

Equilibrium

Neutral equilibrium

Rotation neither raises, nor lowers, the centre of mass.

Tipping will lower the centre of mass.

Unstable equilibrium

Rotation

Centre of mass can be lowered

Object topples if the line of action of the weight is outside the base of the body.

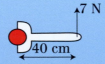

ENERGY Types of Energy and Energy Transfers

Energy is the ability to make something useful happen.

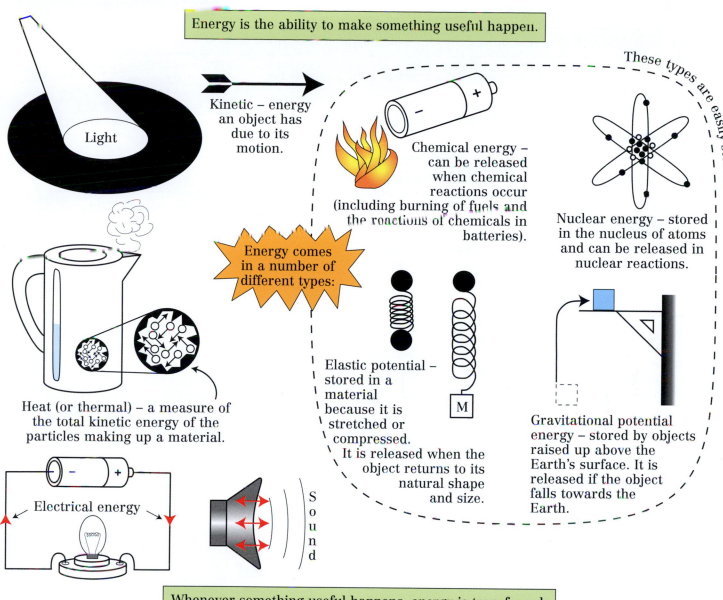

These types are easily stored

Light

Kinetic – energy an object has due to its motion.

Chemical energy – can be released when chemical reactions occur (including burning of fuels and the reactions of chemicals in batteries).

Nuclear energy – stored in the nucleus of atoms and can be released in nuclear reactions.

Energy comes in a number of different types:

Heat (or thermal) – a measure of the total kinetic energy of the particles making up a material.

Elastic potential – stored in a material because it is stretched or compressed. It is released when the object returns to its natural shape and size.

M

Gravitational potential energy – stored by objects raised up above the Earth's surface. It is released if the object falls towards the Earth.

Electrical energy

Sound

Whenever something useful happens, energy is transferred.

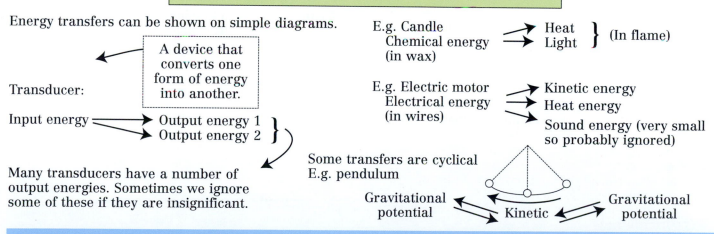

Energy transfers can be shown on simple diagrams.

Transducer:

A device that converts one form of energy into another.

Input energy ⟶ Output energy 1 / Output energy 2

Many transducers have a number of output energies. Sometimes we ignore some of these if they are insignificant.

E.g. Candle
Chemical energy (in wax) ⟶ Heat / Light } (In flame)

E.g. Electric motor
Electrical energy (in wires) ⟶ Kinetic energy / Heat energy / Sound energy (very small so probably ignored)

Some transfers are cyclical
E.g. pendulum

Gravitational potential ⟷ Kinetic ⟷ Gravitational potential

Questions
1. Nuclear energy is stored in the nucleus of atoms. Make a list of the other types of energy that can be stored giving an example of each.
2. What is a transducer? Make a list of five transducers that might be found in a home and the main energy change in each case.
3. Draw an energy transfer diagrams for the following showing the main energy transfers in each case;

   a. Electric filament light bulb.
   b. Solar cell.
   c. Electric kettle.
   d. Loudspeaker.
   e. Mobile 'phone 'charger'.

   f. Clockwork alarm clock.
   g. Playground swing.
   h. Bungee jumper.
   i. Petrol engine.
   j. Microphone.

4. What provides the energy input for the human body? List all types of energy that the body can transfer the energy input into.

ENERGY Energy Conservation

Probably the most important idea in Physics is the Principle of Conservation of Energy, which states:

> Energy cannot be created or destroyed. It can only be transformed from one form to another form.

This means that the total energy input into a process is the same as the total energy output.

We can use a more sophisticated energy transfer diagram, called a *Sankey diagram*, to show this.

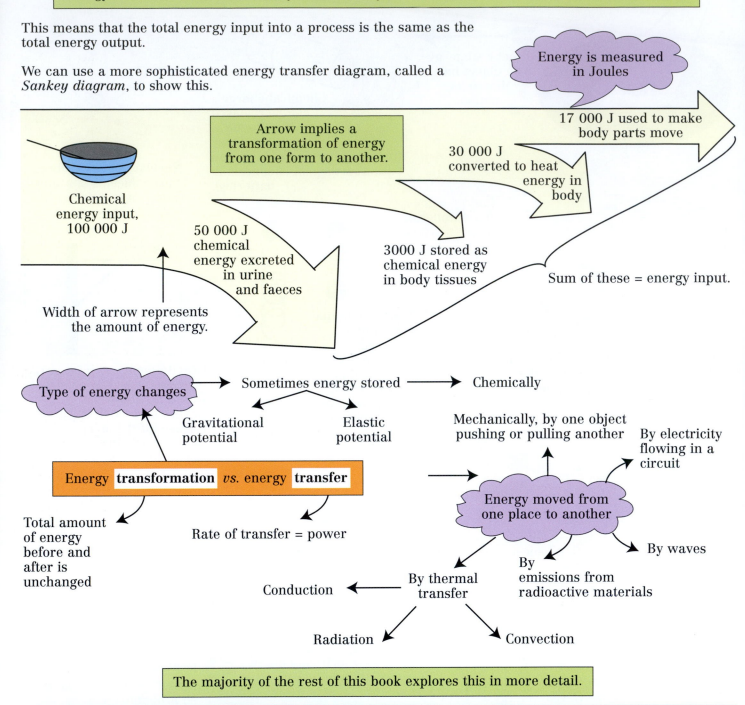

Energy is measured in Joules

17 000 J used to make body parts move

30 000 J converted to heat energy in body

Arrow implies a transformation of energy from one form to another.

Chemical energy input, 100 000 J

50 000 J chemical energy excreted in urine and faeces

3000 J stored as chemical energy in body tissues

Sum of these = energy input.

Width of arrow represents the amount of energy.

Type of energy changes

Sometimes energy stored → Chemically

Gravitational potential   Elastic potential

Mechanically, by one object pushing or pulling another

By electricity flowing in a circuit

Energy **transformation** *vs.* energy **transfer**

Energy moved from one place to another

Total amount of energy before and after is unchanged

Rate of transfer = power

By waves

Conduction ← By thermal transfer

By emissions from radioactive materials

Radiation   Convection

The majority of the rest of this book explores this in more detail.

Questions
1. State the Principle of Conservation of Energy.
2. What units is energy measured in?
3. Explain the difference between energy transformations and energy transfers. Suggest four ways energy can be transferred.
4. A TV set uses 25 J of energy each second. If 15 J of energy is converted to light and 2 J is converted to sound, how much energy is converted to heat, assuming this is the only other form of energy produced?
5. The motor in a toy train produces 1 J of heat energy and 2 J of kinetic energy every second. What must have been the minimum electrical energy input per second? If the train runs uphill and the electrical energy input stays the same, what would happen to its speed?
6. Use the following data to draw a Sankey diagram for each device:
   a. Candle (chemical energy in wax becomes heat energy 80% and light 20%).
   b. Food mixer (electrical energy supplied becomes 50% heat energy in the motor, 40% kinetic energy of the blades, and 10% sound energy).
   c. Jet aircraft (chemical energy in fuel becomes 10% kinetic energy, 20% gravitational potential energy, and 70% heat).

ENERGY Work Done and Energy Transfer

Whenever something useful happens, energy must be transferred but how can we measure energy? The only way to measure energy directly is by considering the idea of *work done*.

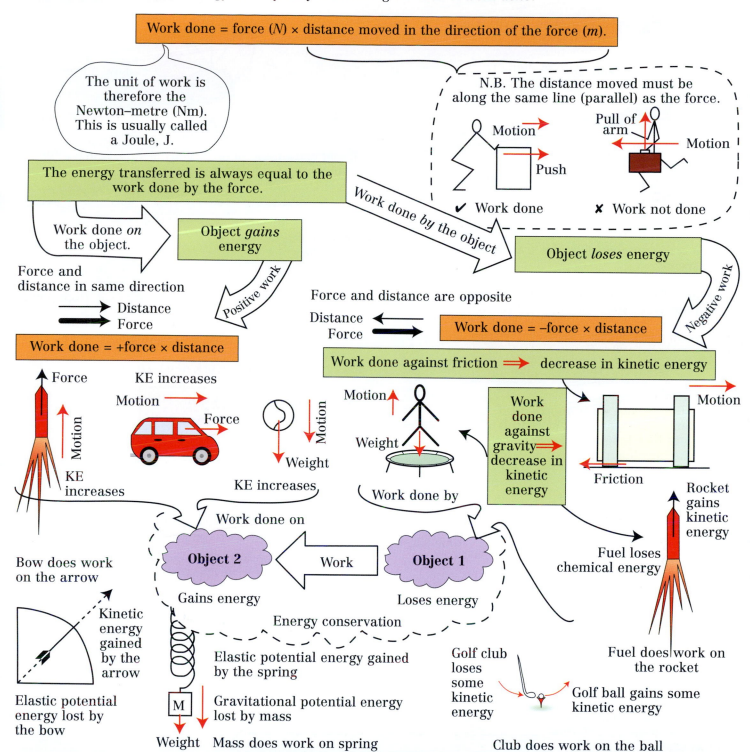

Work done = force (N) × distance moved in the direction of the force (m).

The unit of work is therefore the Newton–metre (Nm). This is usually called a Joule, J.

The energy transferred is always equal to the work done by the force.

Work done on the object.

Object *gains* energy

Work done by the object

N.B. The distance moved must be along the same line (parallel) as the force.

Motion
Push
✔ Work done

Pull of arm
Motion
✗ Work not done

Object *loses* energy

Positive work

Force and distance in same direction

→ Distance
→ Force

Work done = +force × distance

Force
Motion
KE increases

KE increases
Motion
Force

KE increases
Motion
Weight

Force and distance are opposite

Distance
Force

Work done = –force × distance

Negative work

Work done against friction ⟹ decrease in kinetic energy

Motion
Weight

Work done against gravity ⟹ decrease in kinetic energy

Motion
Friction

Motion
Rocket gains kinetic energy

Work done by

Work done on

Bow does work on the arrow

Object 2
Gains energy

Work

Object 1
Loses energy

Energy conservation

Fuel loses chemical energy

Kinetic energy gained by the arrow

Elastic potential energy lost by the bow

Elastic potential energy gained by the spring

M
Gravitational potential energy lost by mass

Weight   Mass does work on spring

Golf club loses some kinetic energy

Golf ball gains some kinetic energy

Club does work on the ball

Fuel does work on the rocket

Questions
1. Copy and complete:
   'Work is done when a ? moves an object. It depends on the size of the ? measured in ? and the ? the object moves measured in ?. Whenever work is done, an equal amount of ? is transferred. The unit of energy is the ?. Work is calculated by the formula: work = ? × distance moved in the ? of the ?.'
2. I push a heavy box 2 m along a rough floor against a frictional force of 20 N. How much work do I do? Where has the energy come from for me to do this work?
3. A parachute exerts a resistive force of 700 N. If I fall 500 m, how much work does the parachute do?
4. A firework rocket produces a constant thrust of 10 N.
   a. The rocket climbs to 150 m high before the fuel is used up. How much work did the chemical energy in the fuel do?
   b. Explain why the chemical energy stored in the fuel would need to be much greater than the work calculated in (a).
   c. The weight of the empty rocket and stick is 2.5 N. How much work has been done against gravity to reach this height?
   d. The answers to parts (a) and (c) are not the same, explain why.

ENERGY Power

Walking

Slow gain in gravitational potential energy.

Low rate of doing work.

Low power

Running

Rapid rate of doing work.

Rapid gain in gravitational potential energy.

High power

Slow conversion of electrical to heat and light energy.

Heat and light energy

40 W

Dim

Low power

Electrical energy

Rapid conversion of electrical to heat and light energy.

Heat and light energy

60 W

Bright

High power

Electrical energy

Power is the number of Joules transferred each second.

The unit of power is the Joule per second, called the Watt, W.

Power is the rate of energy conversion between forms.

Power (W) =	energy transferred (J)
	time taken (s).

'Rate' means how quickly something happens.

Energy transferred = work done, so

Power (W) =	work done (J)
	time taken (s).

Calculating power. Non-mechanical:
- Find out total (heat, light, electrical) energy transferred
- Find out how long the energy transfer took
- Use the formula above

Mechanical:
(i.e. where a force moves through a distance)

Distance

Force

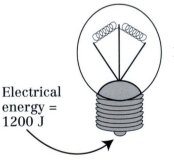

Electrical energy = 1200 J

$$\text{Power} = \frac{\text{energy transferred}}{\text{time taken}}$$

= 1200 J/20 s

= 60 W

- Calculate the work done = force (N) × distance (m)
- Find out how long the work took to be done
- Use the formula above

Bulb is switched on for 20 s.

Compare these: imagine how tired you would get if you personally had to do all the work necessary to generate all the electrical power your house uses.

0.5 m

300 N
20 lifts in 60 s

Work done = 300 N × 0.5 m
= 150 J per lift

Total work done = 20 × 150 J
= 3000 J

$$\text{Power} = \frac{\text{work done}}{\text{time taken}} = \frac{3000 \text{ J}}{60 \text{ s}}$$

= 50 W

Questions

1. A kettle converts 62,000 J of electrical energy into heat energy in 50 s. Show its power output is about 1,200 W.
2. A car travels at constant velocity by exerting a force of 1,025 N on the road. It travels 500 m in 17 s. Show that its power output is about 30 kW.
3. The power to three electrical devices is as follows: energy efficient light bulb, 16 W; the equivalent filament bulb, 60 W; a TV on standby, 1.5 W.
   a. How many more Joules of electrical energy does the filament bulb use in one hour compared to the energy efficient bulb?
   b. Which uses more energy, a TV on standby for 24 hours or the energy efficient bulb on for 1.5 hours?
4. When I bring my shopping home, I carry two bags, each weighing 50 N up two flights of stairs, each of total vertical height 3.2 m. I have a weight of 700 N.
   a. How much work do I do on the shopping?
   b. How much work do I do to raise my body up the two flights of stairs?
   c. If it takes me 30 s to climb all the stairs, show that my power output is about 170 W.

ENERGY Gravitational Potential Energy and Kinetic Energy

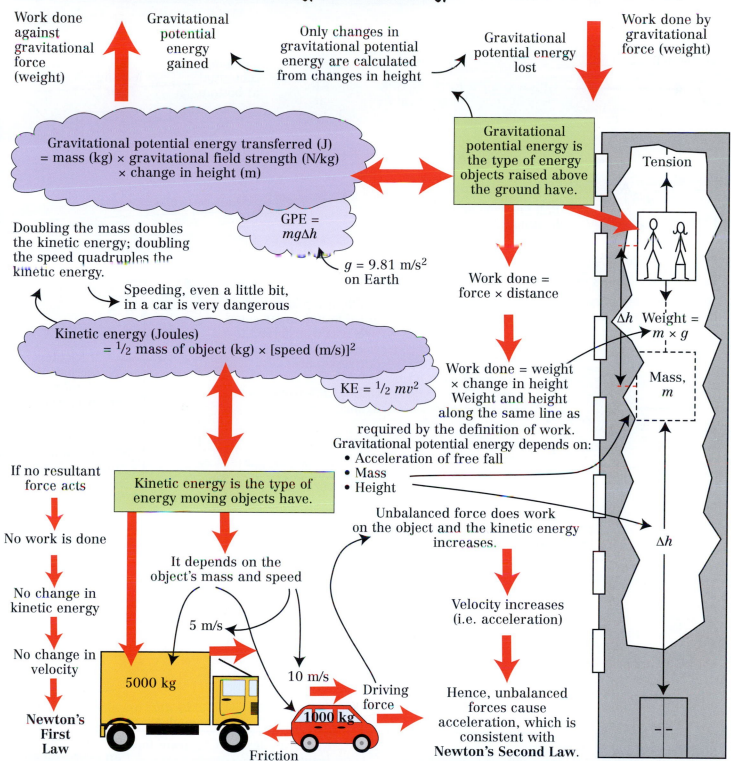

Work done against gravitational force (weight)

Gravitational potential energy gained

Only changes in gravitational potential energy are calculated from changes in height

Gravitational potential energy lost

Work done by gravitational force (weight)

Gravitational potential energy transferred (J) = mass (kg) × gravitational field strength (N/kg) × change in height (m)

Gravitational potential energy is the type of energy objects raised above the ground have.

$$GPE = mg\Delta h$$

Doubling the mass doubles the kinetic energy; doubling the speed quadruples the kinetic energy.

$g = 9.81$ m/s^2 on Earth

Speeding, even a little bit, in a car is very dangerous

Work done = force × distance

Tension

Kinetic energy (Joules) = $^1/_2$ mass of object (kg) × [speed (m/s)]^2

$$KE = ^1/_2\, mv^2$$

Work done = weight × change in height Weight and height along the same line as required by the definition of work.
Gravitational potential energy depends on:
- Acceleration of free fall
- Mass
- Height

Δh Weight = $m \times g$

Mass, m

Kinetic energy is the type of energy moving objects have.

If no resultant force acts

No work is done

No change in kinetic energy

No change in velocity

Newton's First Law

It depends on the object's mass and speed

5 m/s

5000 kg

10 m/s

1000 kg

Driving force

Friction

Unbalanced force does work on the object and the kinetic energy increases.

Δh

Velocity increases (i.e. acceleration)

Hence, unbalanced forces cause acceleration, which is consistent with **Newton's Second Law.**

Questions
1. Make a list of five objects that change their gravitational potential energy.
2. Using the diagram above calculate the kinetic energy of the car and the lorry.
3. How fast would the car have to go to have the same kinetic energy as the lorry?
4. The mass of the lift and the passengers in the diagram is 200 kg. Each floor of the building is 5 m high.
   a. Show that the gravitational potential energy of the lift when on the eighth floor is about 80 000 J.
   b. How much gravitational potential energy would the lift have when on the third floor? If one passenger of mass 70 kg got out on the third floor, how much work would the motor have to do on the lift to raise it to the sixth floor?
   c. What is the gravitational potential energy of a 0.5 kg ball 3 m above the surface of the Moon where the gravitational field strength is about 1.6 N/kg?
5. A coin of mass 10 g is dropped from 276 m up the Eiffel tower.
   a. How much gravitational potential energy would it have to lose before it hits the ground?
   b. Assuming all the lost gravitational potential energy becomes kinetic energy, how fast would it be moving when it hit the ground?
   c. In reality, it would be moving a lot slower, why?

ENERGY Energy Calculations

All energy calculations use the *Principle of Conservation of Energy.*

GPE = gravitational potential energy   KE = kinetic energy

E.g. Bouncing ball

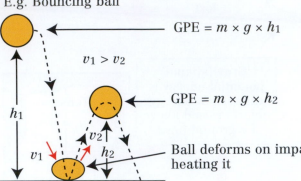

GPE = $m \times g \times h_1$

$v_1 > v_2$

GPE = $m \times g \times h_2$

h_1

v_2

h_2

v_1

Ball deforms on impact, heating it

KE leaving floor $\frac{1}{2}mv_2^2 = mgh_2$ = GPE at h_2

KE hitting floor $\frac{1}{2}mv_1^2 = mgh_1$ = GPE at h_1

Air resistance is ignored

$$\text{GPE at } h_1 = \text{KE at bottom} \quad \begin{array}{c}\text{Elastic potential at bottom}\end{array} = \begin{array}{c}\text{KE leaving floor}\end{array} = \begin{array}{c}\text{GPE at } h_2\end{array}$$

Thermal energy in deformed ball

Conservation of energy
GPE at top of bounce = KE at bottom of bounce
$mg\Delta h = \frac{1}{2}mv_1^2$
$v_1 = \sqrt{2gh}$

GPE on leaving plane = $mg\Delta h$

GPE at top is not equal to KE at bottom as some GPE was transferred to work against friction (air resistance).

GPE = KE + work against friction

$mg\Delta h = \frac{1}{2}mv^2 + F \times \Delta h$

$F = \dfrac{mg\Delta h - \frac{1}{2}mv^2}{\Delta h}$

F

Air resistance, F

If KE stops increasing, *terminal velocity* has been reached.

Δh

W

At terminal velocity, all the loss in GPE is doing work against air resistance.

KE at bottom = $\frac{1}{2}mv^2$

Work against friction

As small as possible to prevent injury

GPE at top of skydive   KE at bottom

Roller Coaster

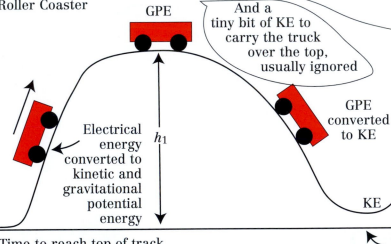

GPE

And a tiny bit of KE to carry the truck over the top, usually ignored

GPE converted to KE

GPE + KE

h_1

Electrical energy converted to kinetic and gravitational potential energy

h_2

KE

$h_2 < h_1$ to ensure truck has enough KE to go over the summit

Time to reach top of track
= GPE gain / power of motor = mgh_1 / power
The time will be greater than this as some electrical energy is converted to KE and does work against friction.

KE here = loss of GPE from top
$\frac{1}{2}mv^2 = mgh_1$
$v = \sqrt{2gh_1}$

This is an overestimate as the truck did work against friction.

Questions Take $g = 9.8 \text{m/s}^2$.

1. At the start of a squash game, a 44 g ball is struck by a racquet and hits the wall at 10 m/s.
   a. Show its KE is about 2 J.
   b. The ball rebounds at 8 m/s. Calculate the loss in KE.
   c. Where, and into what form, has this energy been transferred?
2. An acrobatics aircraft of mass 1000 kg is stationary on a runway. Its take off speed is 150 m/s.
   a. Show that the KE of the aircraft at take off is about 11×10^6 J

b. The maximum thrust of the engines is 20 000 N. Show the aircraft travels over 500 m along the runway before it lifts off.
c. Give two reasons why the runway will actually need to be considerably longer.
d. The aircraft climbs to a height of 1000 m. Show it gains about 10×10^6 J.
e. If the aircraft takes 5 minutes to reach this height, show the minimum power of the engine must be about 33 kW.

f. Why must this be the minimum power?
g. The aircraft then flies level at 200 m/s. What is its KE now?
h. The pilot cuts the engine and goes into a vertical dive as part of the display. When the plane has dived 500 m what is the maximum KE the plane could have gained?
i. Hence, what is the maximum speed the plane could now be travelling at?
j. In reality, it will be travelling slower, why?

26

ENERGY Efficiency and the Dissipation of Energy

If energy is conserved, why do we talk about 'wasting energy'?

Usually when energy is transferred only a proportion of the energy is converted to a useful form, the remainder is converted to other less useful forms of energy, often heat.

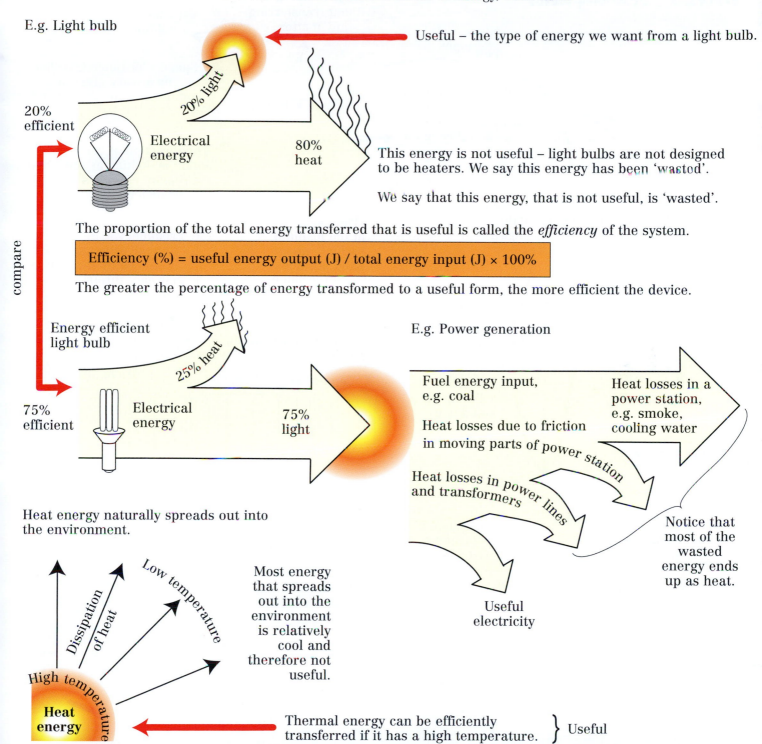

E.g. Light bulb

Useful – the type of energy we want from a light bulb.

20% efficient

20% light

Electrical energy

80% heat

This energy is not useful – light bulbs are not designed to be heaters. We say this energy has been 'wasted'.

We say that this energy, that is not useful, is 'wasted'.

The proportion of the total energy transferred that is useful is called the *efficiency* of the system.

Efficiency (%) = useful energy output (J) / total energy input (J) × 100%

The greater the percentage of energy transformed to a useful form, the more efficient the device.

Energy efficient light bulb

25% heat

75% efficient

Electrical energy

75% light

E.g. Power generation

Fuel energy input, e.g. coal

Heat losses due to friction in moving parts of power station

Heat losses in power lines and transformers

Heat losses in a power station, e.g. smoke, cooling water

Notice that most of the wasted energy ends up as heat.

Useful electricity

Heat energy naturally spreads out into the environment.

Dissipation of heat

Low temperature

High temperature

Heat energy

Most energy that spreads out into the environment is relatively cool and therefore not useful.

Thermal energy can be efficiently transferred if it has a high temperature. } Useful

Questions

1. An electric motor on a crane raises 50 kg of bricks 10 m. If the energy supplied to the motor was 16 000 J show that the motor is about 30% efficient.

2. A rollercoaster has 250 000 J of GPE at the top of the first hill. At the bottom of the first hill, the coaster has 220 000 J of KE. Where did the rest of the energy go, and what is the overall efficiency of the GPE to KE conversion?

3. A ball of mass 30 g falls from 1.5 m and rebounds to 0.8 m. Show that the efficiency of the energy transformation is about 50%. Why do you not need to know the mass of the ball?

4. A car engine is about 20% efficient at converting chemical energy in petrol. If a car of mass 1000 kg has to climb a hill 50 m high, how much chemical energy will be required? Why in reality would substantially more chemical energy be needed than the value you calculated?

5. A filament light bulb produces a lot of waste heat. Explain why this waste heat energy cannot be put to other uses very easily.

6. What are the main sources of energy wastage in:
   a. A vacuum cleaner?
   b. A motor car?

TRANSFER OF ENERGY
WAVES   Describing Waves

All waves transfer energy from one place to another, without transferring any matter.

A wave is a periodic disturbance of a medium.

WAVES

Speed = distance travelled by a wave crest or compression in one second.

The direction of wave motion is defined as the direction energy is transferred.

Two types

Transverse waves

Longitudinal waves

The medium is the material that is disturbed as the wave passes through it.

Particles of the medium oscillate about fixed positions at right angles to the direction of wave travel.

The particles of the medium oscillate about fixed positions along the same line as the wave energy travels.

Frequency is the number of waves per second produced by the source that pass through a given point in the medium. Measured in waves per second or Hertz, Hz.

All particles moving up

Crest (peak)

Wave direction

Particles oscillating up and down

λ

Trough

λ

All particles moving down

Amplitude – distance between a crest or trough and the *undisturbed position*.

Wavelength (λ) – distance between the same point on two adjacent disturbances. Measured in metres.

Particles oscillating side to side

Compression   Rarefaction

Wave direction

λ

Examples longitudinal:
• Sound

Examples transverse:
• Surface water waves
• Light
• Plucked guitar string

Wavefront

Ray at right angles to wavefront

Particles spread out – rarefaction

Particles close – compression

Shows direction of energy transfer

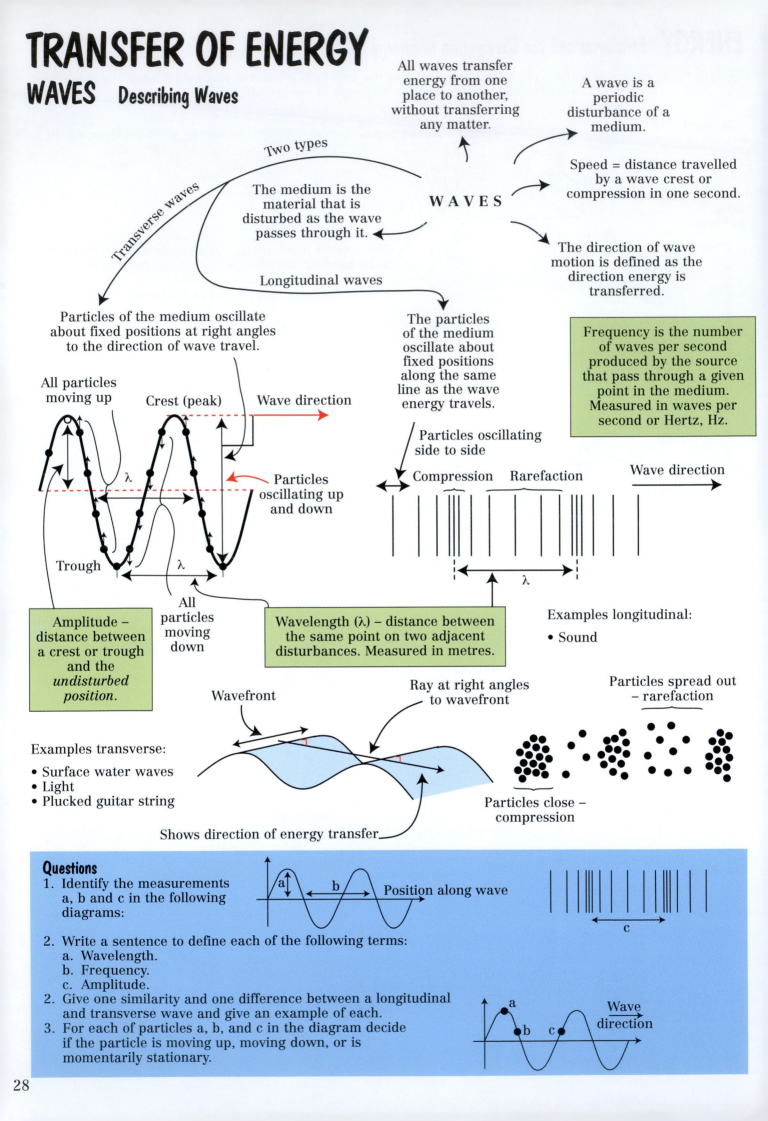

WAVES Wave Speed

The speed of a wave is given by the equation

Wave speed (m/s) = frequency (Hz) × wavelength (m).

Here is how to see why

Walking speed (m/s)   =   stride length (m)   ×   no of steps per second

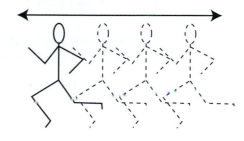

Wave speed (m/s)   =   wavelength (m)   ×   no of waves per second (frequency)

Examples

Water Wave:

frequency = $\dfrac{\text{speed}}{\text{wavelength}}$

$= \dfrac{5 \text{ m/s}}{2 \text{ m}}$

$= 2.5$ Hz

Wave speeds can also be calculated by

Wave speed (m/s)	=	$\dfrac{\text{distance travelled (m)}}{\text{time taken (s)}}$

2 m

5 m/s

E.g. Sonar

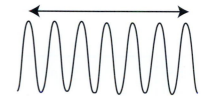

N.B. Remember time is the out and back time.

Light Wave:

Speed of
light = 3×10^8 m/s
frequency = 5×10^{14} Hz,

wavelength = $\dfrac{\text{speed}}{\text{frequency}}$ = $\dfrac{3 \times 10^8 \text{ m/s}}{5 \times 10^{14} \text{ Hz}}$

$= 6 \times 10^{-7}$ m

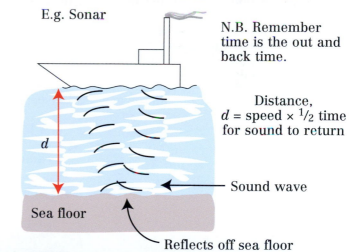

Distance,
d = speed × ½ time
for sound to return

d

Sound wave

Sea floor

Reflects off sea floor

Common speeds:

Speed of light = 3×10^8 m/s (300 000 000 m/s)

Speed of sound ≈ 340 m/s (in air at room temperature)

Questions

1. Calculate the speed of the following waves:
   a. A water wave of wavelength 1 m and frequency 2 Hz.
   b. A water wave of wavelength 3 m and frequency 0.4 Hz.
2. Rearrange the formula wave speed = frequency × wavelength to read:
   a. wavelength = _____.   b. frequency = _____.
3. Calculate the frequency of a sound wave of speed 340 m/s and wavelength:
   a. 2 m.   b. 0.4 m.
4. Calculate the wavelength of a light wave of speed 300 000 000 m/s and frequency:
   a. 4.62×10^{14} Hz.   b. 8.10×10^{14} Hz.

5. Calculate the speed of the following waves. Why might we say that all of these waves belong to the same family?
   a. Wavelength 10 m, frequency = 3×10^7 Hz.
   b. Wavelength 4×10^{-3} m, frequency 7.5×10^{10} Hz.
   c. Wavelength 6×10^{-10} m, frequency 5×10^{17} Hz.
6. In the sonar example above, the echo takes 0.3 s to return from the sea floor. If the sea is 225 m deep, show that the speed of sound in seawater is about 1500 m/s.
7. A radar station sends out radiowaves of wavelength 50 cm and frequency 6×10^8 Hz. They reflect off an aircraft and return in 4.7×10^{-5} s. Show that the aircraft is about 7 km from the radar transmitter.

WAVES Electromagnetic Waves

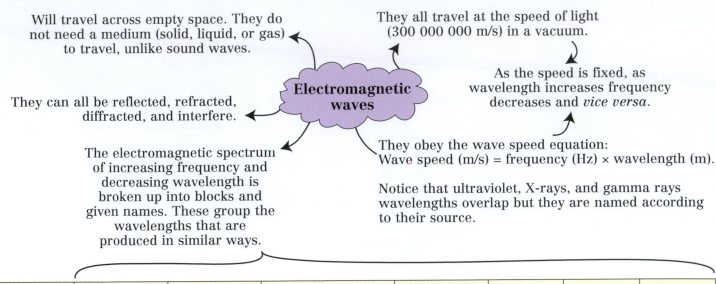

Electromagnetic waves, like all waves transfer energy. They also have the following properties in common.

Will travel across empty space. They do not need a medium (solid, liquid, or gas) to travel, unlike sound waves.

They all travel at the speed of light (300 000 000 m/s) in a vacuum.

Electromagnetic waves

As the speed is fixed, as wavelength increases frequency decreases and *vice versa*.

They can all be reflected, refracted, diffracted, and interfere.

They obey the wave speed equation:
Wave speed (m/s) = frequency (Hz) × wavelength (m).

The electromagnetic spectrum of increasing frequency and decreasing wavelength is broken up into blocks and given names. These group the wavelengths that are produced in similar ways.

Notice that ultraviolet, X-rays, and gamma rays wavelengths overlap but they are named according to their source.

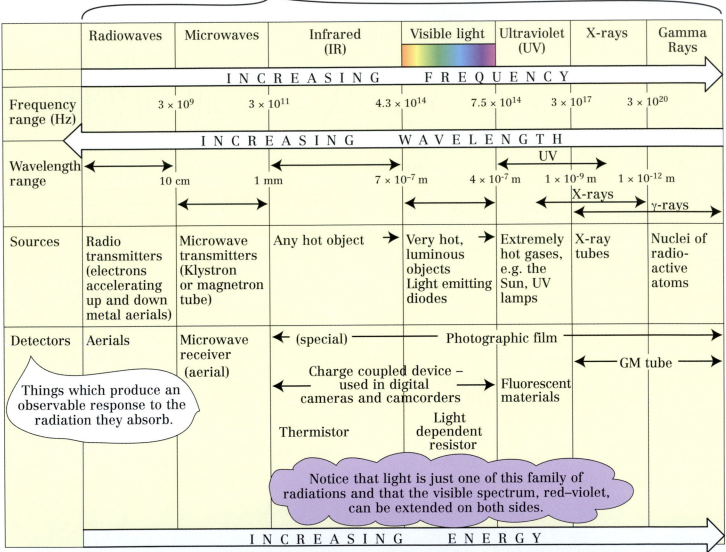

	Radiowaves	Microwaves	Infrared (IR)	Visible light	Ultraviolet (UV)	X-rays	Gamma Rays
INCREASING FREQUENCY →							
Frequency range (Hz)	3×10^9	3×10^{11}	4.3×10^{14}	7.5×10^{14}	3×10^{17}	3×10^{20}	
← INCREASING WAVELENGTH							
Wavelength range	←→ 10 cm	1 mm	7×10^{-7} m	4×10^{-7} m	UV 1×10^{-9} m	X-rays 1×10^{-12} m	γ-rays
Sources	Radio transmitters (electrons accelerating up and down metal aerials)	Microwave transmitters (Klystron or magnetron tube)	Any hot object	Very hot, luminous objects Light emitting diodes	Extremely hot gases, e.g. the Sun, UV lamps	X-ray tubes	Nuclei of radio-active atoms
Detectors	Aerials	Microwave receiver (aerial)	← (special) — Photographic film →			← GM tube →	
			Charge coupled device – used in digital cameras and camcorders		Fluorescent materials		
			Thermistor	Light dependent resistor			

Things which produce an observable response to the radiation they absorb.

Notice that light is just one of this family of radiations and that the visible spectrum, red–violet, can be extended on both sides.

INCREASING ENERGY →

Questions
1. State three properties all electromagnetic waves have in common.
2. Calculate the wavelength of electromagnetic waves of the following frequencies:
   a. 5×10^9 Hz. b. 5×10^{14} Hz. c. 5×10^{15} Hz.
   d. What part of the electromagnetic spectrum does each of these waves come from?
2. Calculate the frequencies of electromagnetic waves of the following wavelengths:
   a. 1 m. b. 1×10^{-5} m. c. 5×10^{-8} m.
   d. What part of the electromagnetic spectrum does each of these waves come from?
3. List the electromagnetic spectrum in order of increasing energy.
4. Which has the longest wavelength, red or blue light? List the colours of the visible spectrum in order of increasing frequency.

WAVES   How Electromagnetic Waves Travel

What is 'waving' in an electromagnetic wave?
It is formed from linked oscillating electric and magnetic fields, hence the name.

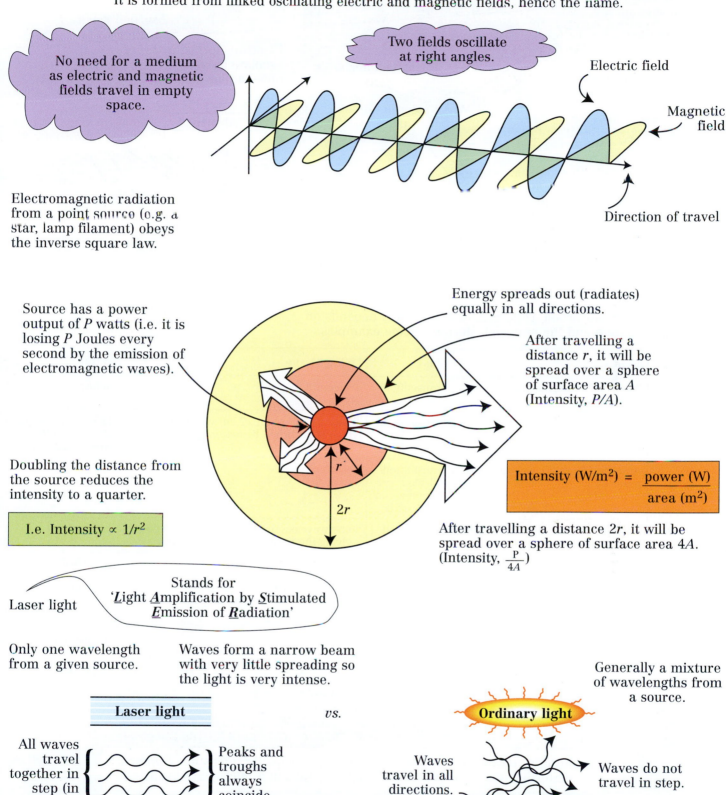

No need for a medium as electric and magnetic fields travel in empty space.

Two fields oscillate at right angles.

Electric field

Magnetic field

Direction of travel

Electromagnetic radiation from a point source (e.g. a star, lamp filament) obeys the inverse square law.

Source has a power output of P watts (i.e. it is losing P Joules every second by the emission of electromagnetic waves).

Energy spreads out (radiates) equally in all directions.

After travelling a distance r, it will be spread over a sphere of surface area A (Intensity, P/A).

Doubling the distance from the source reduces the intensity to a quarter.

I.e. Intensity $\propto 1/r^2$

$$\text{Intensity (W/m}^2) = \frac{\text{power (W)}}{\text{area (m}^2)}$$

After travelling a distance $2r$, it will be spread over a sphere of surface area $4A$. (Intensity, $\frac{P}{4A}$)

Laser light

Stands for 'Light Amplification by Stimulated Emission of Radiation'

Only one wavelength from a given source.

Waves form a narrow beam with very little spreading so the light is very intense.

Generally a mixture of wavelengths from a source.

Laser light

vs.

Ordinary light

All waves travel together in step (in phase).

Peaks and troughs always coincide.

Waves travel in all directions.

Waves do not travel in step.

Questions
1. What is waving in an electromagnetic wave?
2. A 60 W light bulb can be considered a point source of light. What is the intensity of the light:
   a. 1 m from the bulb when it has spread through a sphere of area 12.6 m²?
   b. 2 m from the bulb when it has spread through a sphere of area 50.3 m²?
   c. Suggest what the intensity would be 3 m from the bulb.
3. The intensity of the Sun's radiation at the Earth is about 1400 W/m². Jupiter is about five times further from the Sun. Show that the intensity of the Sun's radiation here is about 56 W/m².
4. Suggest three differences between laser light and ordinary light from a lamp.

WAVES Absorption, Reflection, and Transmission of Electromagnetic Waves

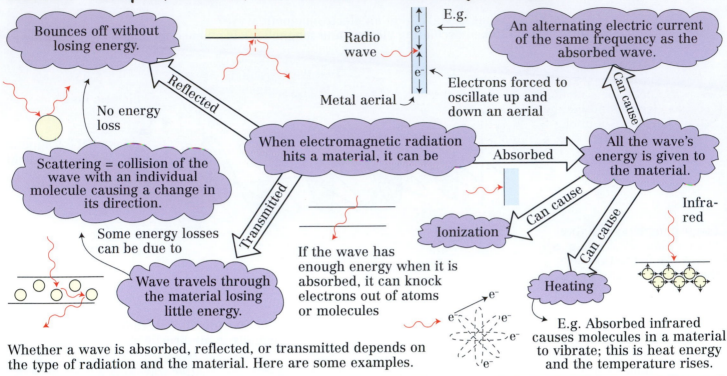

Whether a wave is absorbed, reflected, or transmitted depends on the type of radiation and the material. Here are some examples.

Radiation	Metals	Glass	Living Tissue	Water
Radiowaves	Absorbed by aerials, but otherwise reflected	Transmitted	Transmitted	Reflected
Microwaves	Reflected, e.g. satellite dishes and inside of microwave ovens	Transmitted	Transmitted except 12 cm wavelength which is absorbed by water in the tissues	12 cm wavelength absorbed, otherwise transmitted
Infrared	Absorbed by dull/black surfaces, reflected by shiny ones	Transmitted/reflected depending on wavelength	Absorbed	Absorbed
Visible light	Absorbed by dull/black surfaces, reflected by shiny ones	Transmitted	Some wavelengths absorbed, some reflected – giving the tissue a distinctive colour	Transmitted
Ultraviolet	Absorbed	Absorbed	Absorbed and causes ionization	Absorbed
X-rays	Partially absorbed and partially transmitted. The denser the material the more is absorbed		Partially absorbed and partially transmitted. The denser the tissue the more is absorbed	Transmitted
Gamma rays			Transmitted	Transmitted

Questions
1. Define the following and give an example of a type of radiation and material that illustrates each:
   a. Transmission. b. Reflection. c. Absorption.
2. Suggest three possible results of the absorption of electromagnetic radiation by a material.
3. Copy and complete the table using words below (look ahead to p33 and 34 if you need help).

> Sending signals to mobile phones. Cooking. Aerials. Broadcasting. Suntans. Sterilization. Medical X-rays. Mirrors. Walls of a microwave oven.

	Transmission	Absorption	Reflection
Radiowave			Round the globe broadcasting by bouncing off the ionosphere
Microwave			
Infrared		Cooking	
Visible light	Lenses		
Ultraviolet			
X-ray			
Gamma ray			

WAVES The Earth's Atmosphere and Electromagnetic Radiation

Electromagnetic waves either

pass straight through the atmosphere

or

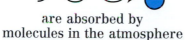

are absorbed by molecules in the atmosphere

or

are scattered by molecules in the atmosphere

Type of radiation	Effect of the atmosphere	Potential uses	Potential problems
Radiowaves	Generally pass straight through, except some long wavelengths will be reflected by a layer called the ionosphere, high in the atmosphere	Carrying messages over long distances. Bouncing radiowaves off the ionosphere allows them to reach receivers out of the line of sight	*(diagram: Ionosphere, Earth)*
Microwaves	Pass through all parts of the atmosphere	Send information to and from satellites in orbit; send information to and from mobile phones; radar	*(diagram: Earth)*
Infrared	Absorbed by water vapour and other gases such as carbon dioxide (present in small amounts) and methane (present in minute amounts)	Humans are increasing the amount of greenhouse gases in the atmosphere. Some scientists think this is causing the Earth to warm up. Possible consequences are . . . • Rising sea levels due to melting of the polar ice caps • Extreme weather conditions occurring more often • Loss of farmland (too wet, dry)	Infrared is emitted by all warm surfaces including the Earth's surface. Some is lost into space but some is absorbed by gases (water, carbon dioxide) in the atmosphere warming it. This is called the *Greenhouse effect* and those gases that absorb infrared, greenhouse gases. Too high a concentration of greenhouse gases leads to *global warming*
Visible light	Passes through clear skies. Blue light is scattered more than red light giving blue skies during the day and red skies at dawn and dusk. *Sunlight / Scattered.* Randomly scattered from water vapour in clouds	Provides plants with energy for photosynthesis and hence all living things with food. Warms the Earth's surface	*(diagram: Molecules in atmosphere, Red, Sunlight, Blue, Evening (sun low in the sky), Earth, Midday (sun overhead))*
Ultraviolet	Absorbed by ozone gas high in the atmosphere (the ozone layer)	Ozone layer protects plants and animals from exposure to too much ionizing ultraviolet radiation from the Sun which would harm them	Ozone layer is being destroyed by chemical reactions with man-made gases
X-rays and gamma rays	Absorbed by the atmosphere		

Questions
1. Which types of electromagnetic radiation pass straight through the atmosphere, which are scattered, and which are absorbed?
2. What is the Greenhouse effect? Suggest why the concentration of carbon dioxide in the atmosphere has been rising for the last 200 years. Suggest three consequences of global warming.
3. Why are cloudy nights generally warmer than when there are clear skies?
4. If the polar ice caps melt, will the Earth's surface absorb more or less radiation from the Sun? Hence will this increase or decrease the rate of global warming?
5. How is the ozone layer helpful to humans and why should we be concerned about a hole in it?

WAVES Uses of Electromagnetic Waves, Including Laser Light

There is an almost limitless range of uses for electromagnetic waves.
The selection below gives a flavour of some of the more common.

Type of radiation	
Radiowaves	Broadcasting (long, medium, and shortwave radio, TV [UHF]) (see pages 97, 99). Emergency services communications
Microwaves	Microwaves are strongly absorbed by water molecules making them vibrate violently. This can be used to heat materials (e.g. food) containing water. Microwave energy penetrates more deeply than infrared so food cooks more quickly Microwaves bounce off the metal walls until absorbed by the food Food must be rotated to ensure all parts are cooked evenly Sending signals to and from mobile phones or orbiting satellites (see p97)
Infrared	Fibre-optic cables (see p104) Remote controls Toasters and ovens Infrared cameras for looking at heat loss from buildings, night vision, and searching for trapped people under collapsed buildings
Visible light	Seeing and lighting Laser light Fibre-optic cables (see p104) To read CDs, DVDs, and barcodes in shops (see p107) Surveying, as laser beams are perfectly straight Eye surgery (can be used to 'weld' a detached retina back into place on the back of the eyeball) Retina
Ultraviolet	Can be produced by passing electrical current through mercury vapour If the tube is coated with a fluorescent chemical this absorbs the ultraviolet radiation and emits visible light Mercury vapour Electric current Washing powder contains fluorescent chemicals to make clothes look 'whiter than white' Fluorescent strip lights Ultraviolet radiation produced Security markers use fluorescent chemicals, which glow in ultraviolet radiation but are invisible in visible light Used for tanning lamps in sun beds
X-rays	Absorption depends on density of the material so can be used to take shadow picture of bones in bodies or objects in luggage (see p108)
Gamma rays	Used to kill cancerous cells Sterilize hospital equipment and food

Questions
1. Write a list of all the things you use electromagnetic radiation for during a typical day.
2. Food becomes hot when the molecules in it vibrate violently. Suggest one similarity and one difference between how this is achieved in a microwave oven and in a conventional thermal oven.
3. Group the uses listed in (1) under the headings:
   a. 'Electromagnetic waves used to communicate'.
   b. 'Electromagnetic waves used to cause a change in a material'.
   c. 'Electromagnetic waves used to gather information'.

WAVES Dangers of Electromagnetic Waves

When electromagnetic radiation is absorbed by the body, it deposits its energy. The more energy deposited, the greater the potential for damage. This depends on the type of radiation, its intensity, and time for which the body is exposed to it.

To reduce the hazard from electromagnetic waves you can reduce the time of exposure, reduce the intensity (for example by moving away from the source or using a lower power source), or by the use of a physical barrier to absorb the radiation.

Type of radiation	Hazard	How to reduce hazard
Non-ionizing. These are a lower hazard		
Radiowaves	Minimal. These generally pass straight through the body and carry little energy	
Microwaves	Low intensity radiation from mobile phones and their transmitter masts may be a health risk, but the evidence is inconclusive	*Reduce time of exposure:* reduce phone usage *Reduce intensity:* use a hands free kit to reduce exposure
	Microwaves used in ovens causes a heating effect in water, which would therefore damage water-containing cells	*Physical barrier:* microwave ovens have metal case and grille over the door to prevent microwaves escaping
Infrared	Absorbed infrared can lead to cell damage, which we call a burn	*Reduce time of exposure and intensity:* the body has a natural defence mechanism of instinctively moving away from sources of infrared that are uncomfortably hot
Visible light	Only laser light presents a significant hazard	*Reduce exposure:* never look into the beam
		Physical barrier: most laser products, especially if high intensity, have the beam shielded
Ionizing – able to break molecules into smaller parts (ions) which may go on to be involved in further (possibly harmful) chemical reactions. If these molecules are in the cells of the body the ions can cause changes to the DNA of the cell causing it to divide and grow incorrectly. This is called *cancer*		
Ultraviolet	Absorption may cause cell mutations (particularly in skin) which can lead to cancer Sunburn	*Physical barrier:* sun cream and sun block contain chemicals that strongly absorb ultraviolet providing a barrier between the radiation and the skin Wear clothing *Reduce time of exposure:* avoid excessive sunbathing or tanning treatment
X-rays	Some absorbed and some transmitted. Absorbed radiation may cause cell mutations leading to cancer	*Reduce time of exposure:* limit number of X-rays you are exposed to (but sometimes the medical benefits outweigh the potential risks) *Physical barrier:* health workers use lead shielding to reduce their exposure
Gamma rays	High enough energy to directly kill cells (radiation burns), or to cause cancerous cell mutation	*Physical barrier:* gamma rays from nuclear power plants are shielded from the outside by thick layers of lead, steel, and concrete *Reduce time of exposure:* nuclear industry workers have their exposure times carefully monitored and controlled

Questions
1. Suggest three ways that exposure to harmful electromagnetic waves can be reduced.
2. What is the difference between ionizing and non-ionizing radiation?
3. A parent is worried about the possible health risks of a child using a mobile phone while sunbathing in swimwear on a very sunny day. What advice would you give them?

WAVES Reflection, Refraction and Total Internal Reflection

Waves *reflect* off a plane surface.

Ray shows direction of wave travel

Angle of incidence, *i* = angle of reflection, *r*

Normal – a construction line perpendicular to the reflecting / refracting surface at the point of incidence

If the waves meet the boundary at an angle . . .

Fast waves

Ray shows wave direction.

This part of the wavefront slows down first.

This part of the wavefront continues at a higher speed.

Wave changes direction.

Wave has turned towards the normal as it slows down.

Slow waves

This part of the wavefront speeds up first.

This part continues at a lower speed.

Fast waves

Wave changes direction away from the normal.

Wave now parallel to original wave.

This process is called *refraction*.

Waves travel at different speeds depending on the media they are travelling in.

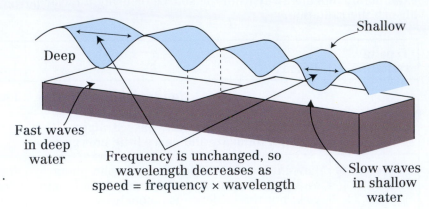

Shallow

Deep

Fast waves in deep water

Frequency is unchanged, so wavelength decreases as speed = frequency × wavelength

Slow waves in shallow water

Think about cars on a road, if they slow down they get closer together but the number of cars passing each second stays the same.

Fast

Slow

The material light passes through is called the *medium*.

If the speed of light is different in two different media, it also refracts.

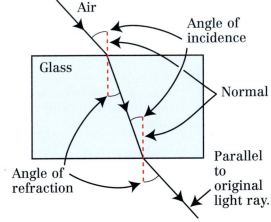

Air

Angle of incidence

Glass

Normal

Angle of refraction

Parallel to original light ray.

As light slows down it changes direction towards the normal (angle of incidence, *i*, > angle of refraction, *r*).

As light speeds up it changes direction away from the normal (angle of incidence, *i*, < angle of refraction, *r*).

Real and apparent depth

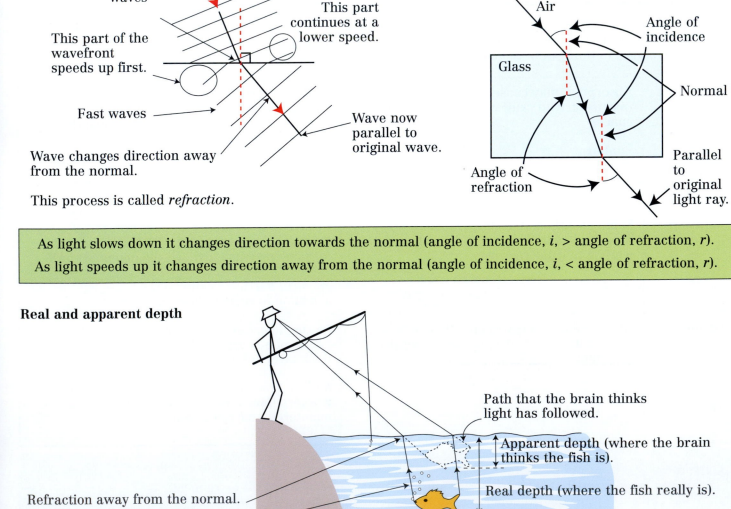

Path that the brain thinks light has followed.

Apparent depth (where the brain thinks the fish is).

Real depth (where the fish really is).

Refraction away from the normal.

Actual path of light ray.

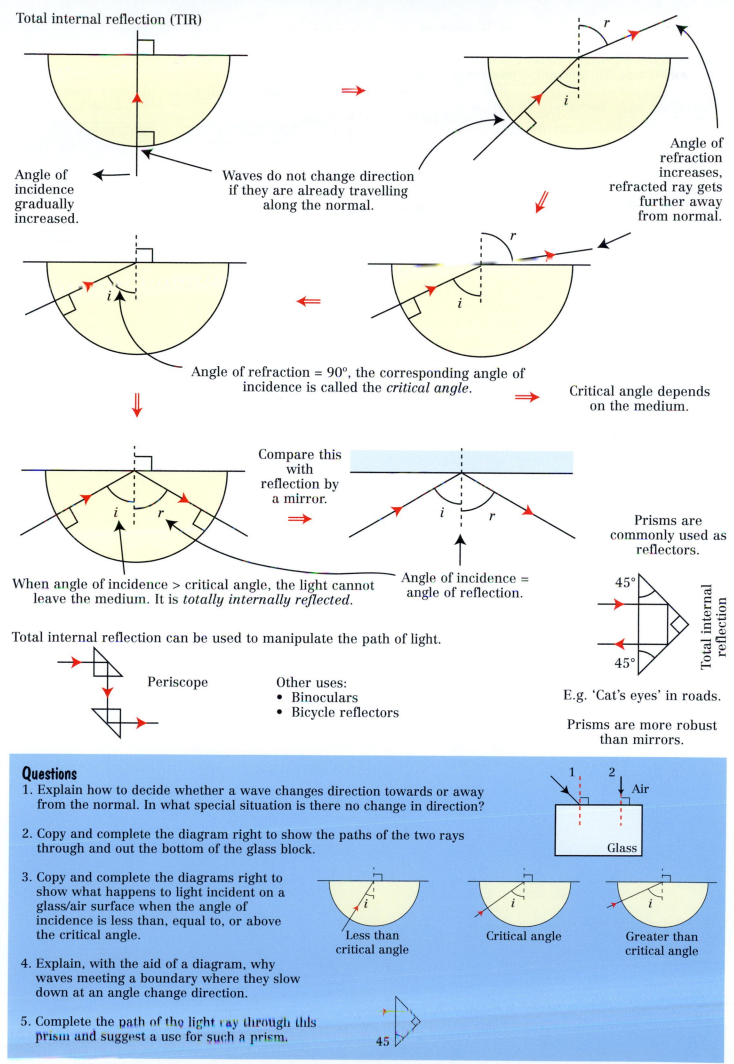

Total internal reflection (TIR)

Waves do not change direction if they are already travelling along the normal.

Angle of incidence gradually increased.

Angle of refraction increases, refracted ray gets further away from normal.

Angle of refraction = 90°, the corresponding angle of incidence is called the *critical angle*.

Critical angle depends on the medium.

Compare this with reflection by a mirror.

When angle of incidence > critical angle, the light cannot leave the medium. It is *totally internally reflected*.

Angle of incidence = angle of reflection.

Prisms are commonly used as reflectors.

45°

45°

Total internal reflection

Total internal reflection can be used to manipulate the path of light.

Periscope

Other uses:
• Binoculars
• Bicycle reflectors

E.g. 'Cat's eyes' in roads.

Prisms are more robust than mirrors.

Questions

1. Explain how to decide whether a wave changes direction towards or away from the normal. In what special situation is there no change in direction?

2. Copy and complete the diagram right to show the paths of the two rays through and out the bottom of the glass block.

   1 2 Air
   Glass

3. Copy and complete the diagrams right to show what happens to light incident on a glass/air surface when the angle of incidence is less than, equal to, or above the critical angle.

   Less than critical angle   Critical angle   Greater than critical angle

4. Explain, with the aid of a diagram, why waves meeting a boundary where they slow down at an angle change direction.

5. Complete the path of the light ray through this prism and suggest a use for such a prism.

   45

WAVES   Refractive Index and Dispersion

When light travels from a vacuum (or air since it makes very little difference to the speed) into another medium, it is slowed down. The amount of slowing is expressed by the ratio:

$$\frac{\text{Speed of light in vacuum (m/s)}}{\text{Speed of light in medium (m/s)}} = \text{refractive index, } n$$

Small refractive index

Large refractive index

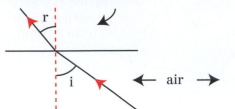

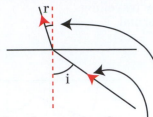

← air →

Therefore, the bigger the refractive index the greater the change in direction of the light wave as it passes into the medium.

Hence Snell's Law

$$\text{Refractive index } n, = \frac{\sin i}{\sin r}$$

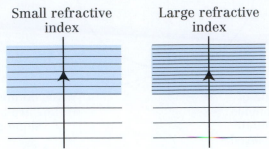

Small refractive index

Large refractive index

The bigger the refractive index the more the light is slowed as it passes into the medium.

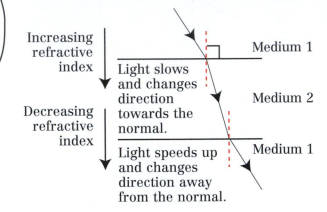

Increasing refractive index

Medium 1

Light slows and changes direction towards the normal.

Medium 2

Decreasing refractive index

Light speeds up and changes direction away from the normal.

Medium 1

Dispersion

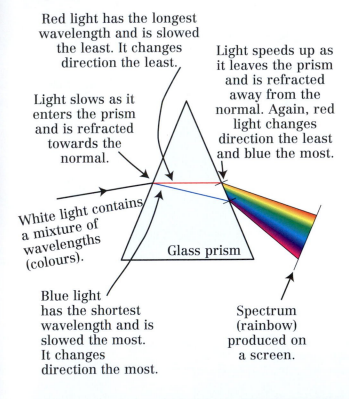

Red light has the longest wavelength and is slowed the least. It changes direction the least.

Light speeds up as it leaves the prism and is refracted away from the normal. Again, red light changes direction the least and blue the most.

Light slows as it enters the prism and is refracted towards the normal.

White light contains a mixture of wavelengths (colours).

Glass prism

Blue light has the shortest wavelength and is slowed the most. It changes direction the most.

Spectrum (rainbow) produced on a screen.

Total internal reflection

1. Light must change direction away from the normal so must be going from high to low refractive index.
2. Angle of incidence must be greater than the critical angle.

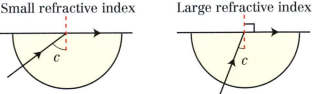

Small refractive index

Large refractive index

The higher the refractive index of the material, the greater the change of direction away from the normal and therefore, the lower its critical angle.

The critical angle, c, can be calculated from the ratio of the refractive indices either side of the boundary.

$$\text{Sin (critical angle)} = \frac{\text{refractive index of second material}}{\text{refractive index of first material}}$$

$$\text{Sin } c = \frac{n_r}{n_i}$$

Questions
1. Which colour, blue or red, is slowed most as it enters a glass prism?
2. Copy the water droplet and complete the diagram to show how the drop splits the white light into colours. Show the order of these colours on your diagram.
3. The speed of light in a vacuum is 3×10^8 m/s. Show that:
   a. The refractive index of water is about 1.3 given the speed of light in water is 2.256×10^8 m/s.
   b. The speed of light in diamond is about 1.2×10^8 m/s given its refractive index is 2.42.
4. The refractive index of glass is about 1.52. A ray of light enters a glass block at 25° to the normal. Show that it continues through the block at about 16°.
5. What is the critical angle for light travelling from water, refractive index 1.33, to air, refractive index 1.00? Why is it not possible to calculate a critical angle for light travelling from air into water?

White light

WAVES Diffraction and Interference

Both diffraction and interference are properties of waves. The fact that all electromagnetic waves display both effects is strong evidence for them having a wave nature.

Diffraction – the spreading out of wave energy as it passes through a gap or past an obstacle.

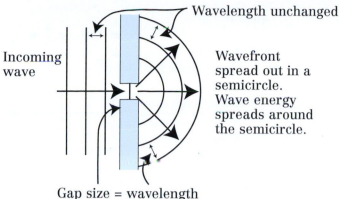

Wavelength unchanged

Incoming wave

Wavefront spread out in a semicircle. Wave energy spreads around the semicircle.

Gap size = wavelength

Wavelength unchanged

Majority of wave energy continues ahead, a small proportion of the energy spreads out.

Gap size much wider than wavelength – diffraction effect is not very noticeable.

Light has a very short wavelength (about 5×10^{-7} m), so needs very small gap sizes for diffraction to be noticeable.

Interference – when two waves meet, their effects add.

When two waves arrive in step, they reinforce each other and this is called *constructive interference*. For light the result would be bright and for sound, loud.

When two waves arrive out of step they cancel out and this is called *destructive interference*. For light this would be dark and for sound, quiet.

Waves set off in step

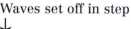

Waves meet here

l_1

l_2

If the difference in path length ($l_1 - l_2$) is:

- A whole number of wavelengths the waves arrive in step and we have constructive interference.
- An odd number of half wavelengths, the waves arrive out of step and we have destructive interference.

Interference patterns

Path difference = whole number of wavelengths, here the waves arrive in step and add

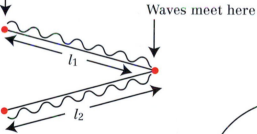

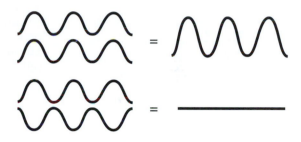

Path difference = odd number of half wavelengths, here the waves arrive out of step and cancel out

Slits

Questions

1. The speed of sound in air is about 340 m/s. Calculate wavelength of the note 'middle C', frequency = 256 Hz. Hence, explain why a piano can be heard through an open doorway, even if the piano itself cannot be seen.
2. A satellite dish behaves like a gap with electromagnetic waves passing through. Explain why the dish sending the signal to a satellite should have a much wider diameter than the wavelength of the waves, whereas a dish broadcasting a signal from a satellite over a wide area should have the same diameter as the wavelength of the waves.
3. The diagram shows a plan view of a harbour. The wavelength of the waves arriving from the sea is 10 m.
   a. How long is length x?
   b. How many waves fit in the length E_1 to B?
   c. How many waves fit in the length E_2 to B?
   d. Therefore, will the waves arrive in or out of step at the buoy, B? Hence, describe the motion of a boat tied to it.
   e. If the wavelength increased to 20 m how would your answers to b–d change?

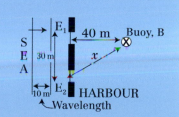

WAVES Polarization and the Photon Model of Light

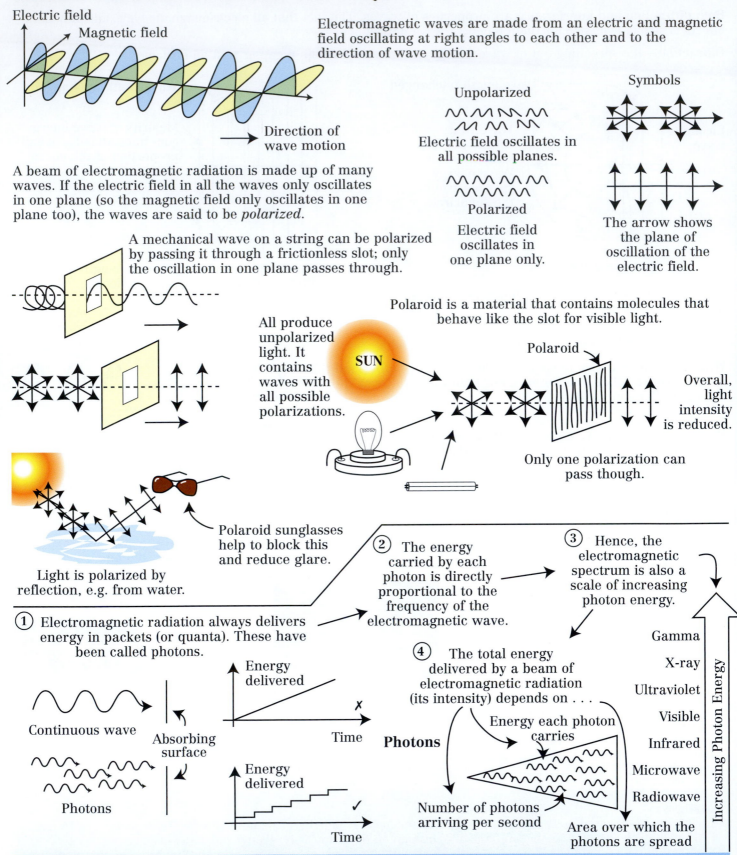

Electric field

Magnetic field

Direction of wave motion

Electromagnetic waves are made from an electric and magnetic field oscillating at right angles to each other and to the direction of wave motion.

A beam of electromagnetic radiation is made up of many waves. If the electric field in all the waves only oscillates in one plane (so the magnetic field only oscillates in one plane too), the waves are said to be *polarized*.

Unpolarized

Electric field oscillates in all possible planes.

Polarized

Electric field oscillates in one plane only.

Symbols

The arrow shows the plane of oscillation of the electric field.

A mechanical wave on a string can be polarized by passing it through a frictionless slot; only the oscillation in one plane passes through.

All produce unpolarized light. It contains waves with all possible polarizations.

SUN

Polaroid is a material that contains molecules that behave like the slot for visible light.

Polaroid

Overall, light intensity is reduced.

Only one polarization can pass though.

Light is polarized by reflection, e.g. from water.

Polaroid sunglasses help to block this and reduce glare.

① Electromagnetic radiation always delivers energy in packets (or quanta). These have been called photons.

Continuous wave

Absorbing surface

Photons

Energy delivered

Time   ✗

Energy delivered

Time   ✓

② The energy carried by each photon is directly proportional to the frequency of the electromagnetic wave.

③ Hence, the electromagnetic spectrum is also a scale of increasing photon energy.

④ The total energy delivered by a beam of electromagnetic radiation (its intensity) depends on . . .

Photons

Energy each photon carries

Number of photons arriving per second

Area over which the photons are spread

Gamma
X-ray
Ultraviolet
Visible
Infrared
Microwave
Radiowave

Increasing Photon Energy

Questions
1. What do we mean by a polarized wave? Draw a diagram to illustrate your answer.
2. Reflected light from a lake in summer is horizontally polarized. Which orientation of light should the Polaroid material in sunglasses allow to pass if the glasses are to cut down glare from the lake?
3. What is a photon?
4. What type of radiation delivers more energy per photon, X-rays or radiowaves?
5. Suggest why X-rays and gamma rays can knock electrons out of atoms (ionize them) but visible light and infrared cannot. What effect might this have on the human body?
6. The photons in a beam of electromagnetic radiation carry 4×10^{-17} J each. If 1×10^{18} photons arrive each second over a 2 m² area what is the total energy arriving per m²?

WAVES   Seismic Waves and the Structure of the Earth

Earthquakes occur when stresses build up at fault lines where the Earth's tectonic plates are moving past each other. The energy stored can be suddenly released as the plates shift, sending out a shock or seismic wave.

Epicentre – the point directly above the focus on the Earth's surface.

Focus – point at which stress between plates is released.

These seismic waves come in two types

P (primary) waves	S (secondary) waves
Faster (about 10 km/s)	Slower (about 6 km/s)
Longitudinal	Transverse
Travel through solids and liquids	Travel through solids only
Cause most damage as they make buildings move side to side	Cause less damage as they make buildings move up and down

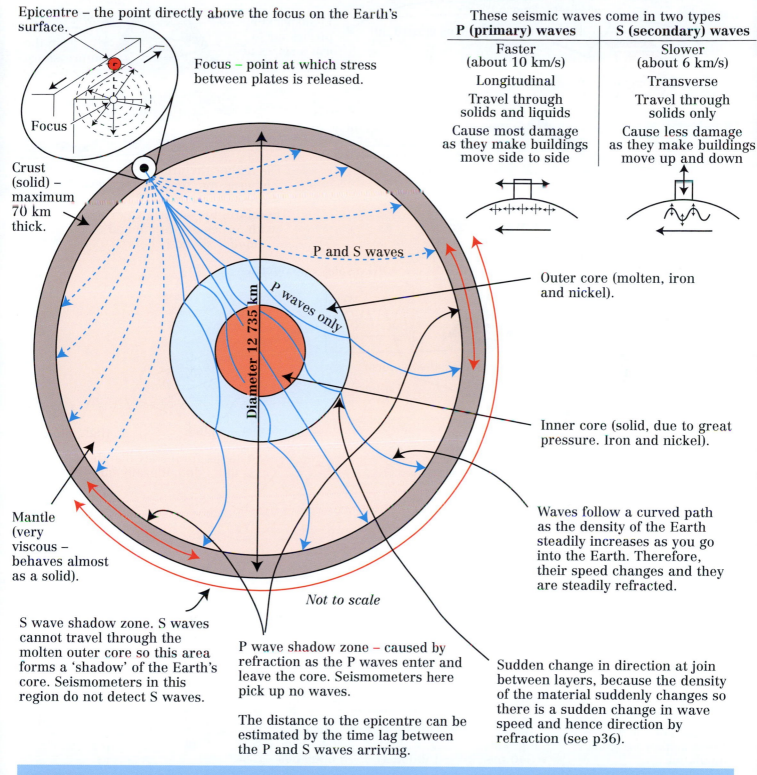

Crust (solid) – maximum 70 km thick.

P and S waves

P waves only

Diameter 12 735 km

Outer core (molten, iron and nickel).

Inner core (solid, due to great pressure. Iron and nickel).

Mantle (very viscous – behaves almost as a solid).

Waves follow a curved path as the density of the Earth steadily increases as you go into the Earth. Therefore, their speed changes and they are steadily refracted.

Not to scale

S wave shadow zone. S waves cannot travel through the molten outer core so this area forms a 'shadow' of the Earth's core. Seismometers in this region do not detect S waves.

P wave shadow zone – caused by refraction as the P waves enter and leave the core. Seismometers here pick up no waves.

The distance to the epicentre can be estimated by the time lag between the P and S waves arriving.

Sudden change in direction at join between layers, because the density of the material suddenly changes so there is a sudden change in wave speed and hence direction by refraction (see p36).

Questions
1. What is the difference between an Earthquake's epicentre and its focus?
2. Draw a labelled diagram of the layers in the Earth. If the crust is a maximum of 70 km thick, what percentage of the total radius of the Earth is made up of crust?
3. Write down two similarities and three differences between P and S waves.
4. Explain how scientists know that the outer core of the Earth is molten.
5. Here is a seismometer trace for an earthquake:
   a. Which trace, X or Y, shows the arrival of the S waves and which the P waves?
   b. If the speed of the P waves is 10 km/s and they took 150 s to arrive, how far away was the earthquake?
   c. If the speed of the S waves is 6 km/s, how long should they take to arrive?
   d. Hence, what is the time interval t marked on the graph?

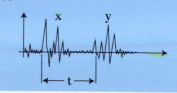

WAVES   Sound Waves

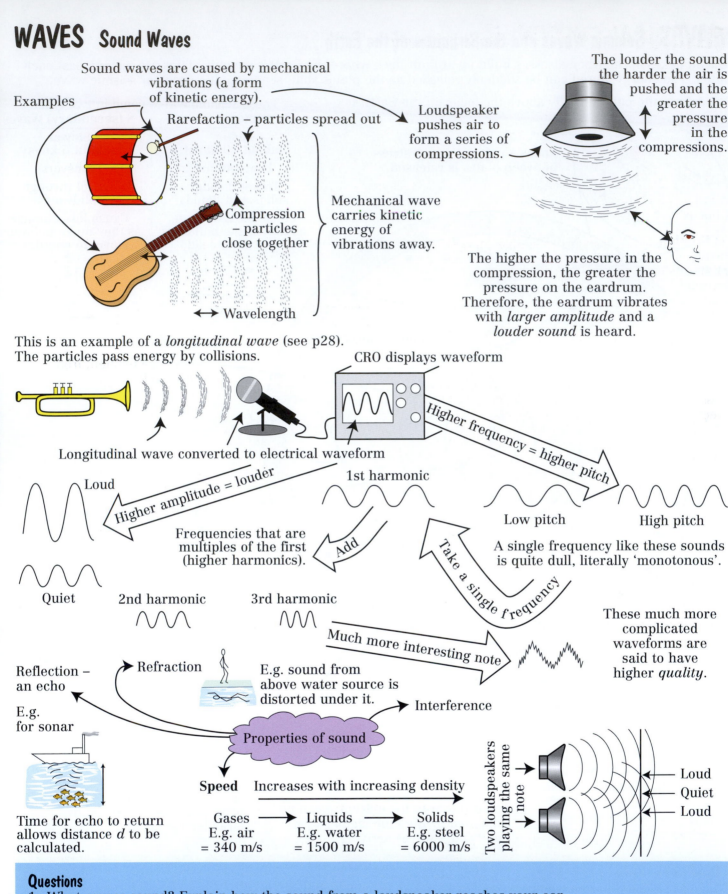

Sound waves are caused by mechanical vibrations (a form of kinetic energy).

Examples

Rarefaction – particles spread out

Loudspeaker pushes air to form a series of compressions.

The louder the sound the harder the air is pushed and the greater the pressure in the compressions.

Compression – particles close together

Mechanical wave carries kinetic energy of vibrations away.

↔ Wavelength

The higher the pressure in the compression, the greater the pressure on the eardrum. Therefore, the eardrum vibrates with *larger amplitude* and a *louder sound* is heard.

This is an example of a *longitudinal wave* (see p28). The particles pass energy by collisions.

CRO displays waveform

Longitudinal wave converted to electrical waveform

Higher frequency = higher pitch

Loud

Higher amplitude = louder

1st harmonic

Low pitch   High pitch

Quiet

Frequencies that are multiples of the first (higher harmonics).

Add

Take a single frequency

A single frequency like these sounds is quite dull, literally 'monotonous'.

2nd harmonic

3rd harmonic

Much more interesting note

These much more complicated waveforms are said to have higher *quality*.

Reflection – an echo

Refraction

E.g. sound from above water source is distorted under it.

Interference

E.g. for sonar

Properties of sound

Time for echo to return allows distance *d* to be calculated.

Speed   Increases with increasing density

Gases	Liquids	Solids
E.g. air = 340 m/s	E.g. water = 1500 m/s	E.g. steel = 6000 m/s

Two loudspeakers playing the same note

Loud
Quiet
Loud

Questions

1. What causes sound? Explain how the sound from a loudspeaker reaches your ear.
2. Explain why sound cannot travel in a vacuum.
3. Use the formula speed = frequency × wavelength to calculate the range of wavelengths of sound the human ear can hear in air where the speed of sound is about 340 m/s.
4. Why does sound travel faster in solids than in gases?
5. What does the pitch of a sound wave depend on?
6. What does the loudness of a sound wave depend on?
7. What is a harmonic?
8. Copy this waveform and add:
   a. A waveform of twice the frequency but the same amplitude.
   b. A waveform of half the amplitude but the same frequency.
   c. A waveform of the same amplitude and frequency but of a higher quality.

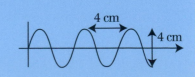

4 cm

4 cm

ELECTRICAL ENERGY Static Electricity

The positively charged nucleus is orbited by negatively charged electrons. These do not escape because opposite charges attract.

All materials are made of atoms

Normally the number of positive and negative charges is equal in each atom

Nucleus – has positive charge

Electrons – have negative charge

Measured in Coulombs, C.

Static electricity is formed when electrical charges are trapped on an insulating material that does not allow them to move. You can charge up a material by . . .

Like charges repel

Unlike charges attract

1) Friction

Wool duster

Electrons (negative charge) rubbed off the wool onto the polythene rod.

Gain of electrons leaves an overall negative charge.

Loss of electrons leaves an overall positive charge.

Polythene rod

In all charging processes, it is always electrons that move as they can be removed from or added to atoms.

Clothing and aircraft can be charged by friction.

Static electricity will flow to Earth if a conducting path is provided, as the charges can get further apart by spreading over the Earth.

FUEL

Both plane and fuel tanker are connected to Earth before fuelling to prevent any sparks.

E.g.

Danger of explosions – flammable vapours or dust ignited by sparks – earthing needed in these environments

2) Induction

Overall neutral particle, e.g. dust.

Electrons repelled away.

Opposite charges attract

Surplus of positive charges on this surface

Negatively charged surface (e.g. a TV screen).

This explains why TV screens tend to get very dusty.

E.g. polish

Antistatic sprays contain a conducting chemical to avoid the build up of charge.

Copper lightning conductor

Sparks like lightning are attracted to sharp points.

Large metal plate

Uses and dangers of static electricity

Spraying paint and pesticides

Negatively charged metal object.

Attraction

Spray gun is positively charged.

As drops all have the same charge, they repel giving a fine spray.

Negative charge induced on leaf by positive drops.

Positively charged droplets are attracted to leaf.

Electrostatic precipitator

Large voltage

Smoke attracted and sticks to negative plates.

Chimney

Negatively charged plate

Positively charged grid.

Smoke particles positively charged by contact with the grid.

When large particles are collected on the plates, they fall down the chimney due to their weight.

Prevents dirty smoke entering the atmosphere.

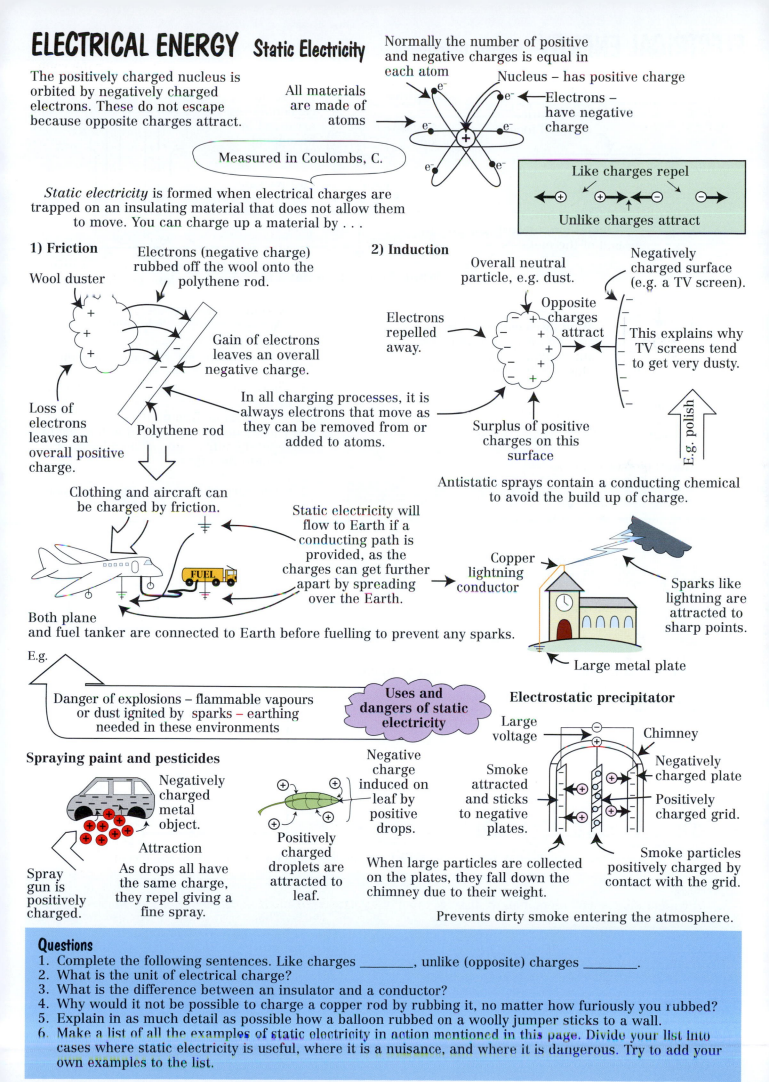

Questions
1. Complete the following sentences. Like charges _____, unlike (opposite) charges _____.
2. What is the unit of electrical charge?
3. What is the difference between an insulator and a conductor?
4. Why would it not be possible to charge a copper rod by rubbing it, no matter how furiously you rubbed?
5. Explain in as much detail as possible how a balloon rubbed on a woolly jumper sticks to a wall.
6. Make a list of all the examples of static electricity in action mentioned in this page. Divide your list into cases where static electricity is useful, where it is a nuisance, and where it is dangerous. Try to add your own examples to the list.

ELECTRICAL ENERGY Electric Currents

An electric current is a flow of charged particles.

Solids

Fixed lattice of metal ions.

Negative

$\oplus$ e→ $\oplus$ e→ $\oplus$ e→
e→ $\oplus$ e→ $\oplus$ e→ $\oplus$

Positive

Metals

Electrons move from metal ion to metal ion, towards the positive end of the metal.

Graphite

Negative side

Layers of carbon atoms arranged in hexagonal sheets.

e⁻

e⁻→ →

Positive side

Electrons move through the layers.

Liquids

Negative ion   Positive ion

Molten or dissolved *ionic* solids.

Gases

Atom

Atom loses electrons due to heating (or by collision with other fast moving atoms). This is called ionization.

Heat

Positive ion

Current flows between plates.

What do all these have in common?

Three conditions for an electric current to flow:
1. There must be charge carriers (electrons or ions).
2. They must be free to move.
3. There must be a potential difference to repel them from one side and attract them to the other.

Electric current is always measured with an ammeter, always placed in series.

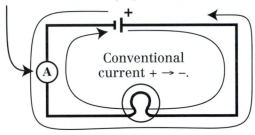

Conventional current + → −.

Electron current − → +.

> Current (in Amps) is the rate of flow of charge; the number of Coulombs of charge flowing past a point per second.
>
> $$\text{Current (A)} = \frac{\text{Charge (C)}}{\text{time (seconds, s)}}$$
>
> 1 Amp = 1 coulomb per second
>
> In equations we usually use I for current and Q for charge. Hence $I = Q/t$.

Current rules
1. The current is the same all the way round a series circuit. Current is **not** used up.

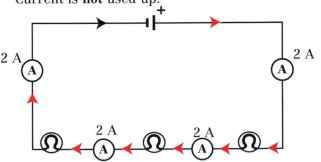

2. The current flowing into a junction = current flowing out.

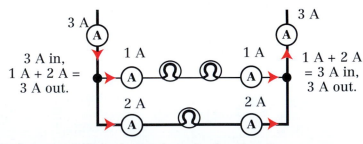

3 A in, 1 A + 2 A = 3 A out.

1 A + 2 A = 3 A in, 3 A out.

These rules mean that charge is conserved. It does not 'pile up' anywhere in the circuit.

Questions
1. Why must ionic solids be molten or dissolved to conduct an electric current?
2. In a circuit 4 C of charge passes through a bulb in 2.5 s. Show that the current is 1.6 A.
3. An ammeter in a circuit shows a current of 1.2 A.
   a. The current flows for 2 minutes. Show the total charge passing through the ammeter is 144 C.
   b. How long would it take 96 C to pass through the ammeter?
4. In the following circuit, how many Amps flow through the battery?
5. The laws of circuit theory were all worked out in the 1800s. The electron was discovered in 1897. Discuss why we have conventional direct current flowing from positive to negative, when we know that the electrons actually flow from negative to positive.

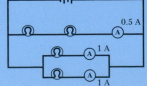

44

ELECTRICAL ENERGY Potential Difference and Electrical Energy

What actually happens in an electric circuit?

We can use a model to help us understand what is happening.

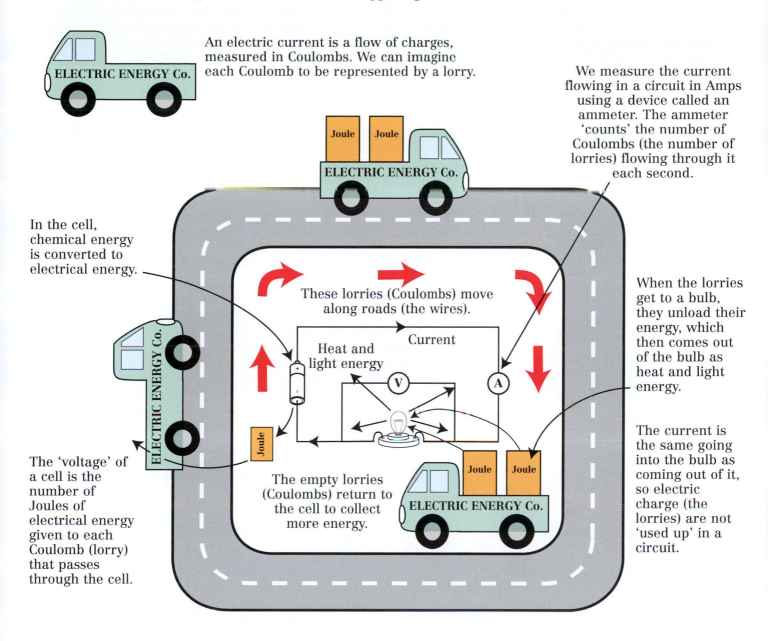

An electric current is a flow of charges, measured in Coulombs. We can imagine each Coulomb to be represented by a lorry.

We measure the current flowing in a circuit in Amps using a device called an ammeter. The ammeter 'counts' the number of Coulombs (the number of lorries) flowing through it each second.

In the cell, chemical energy is converted to electrical energy.

These lorries (Coulombs) move along roads (the wires).

When the lorries get to a bulb, they unload their energy, which then comes out of the bulb as heat and light energy.

The 'voltage' of a cell is the number of Joules of electrical energy given to each Coulomb (lorry) that passes through the cell.

The empty lorries (Coulombs) return to the cell to collect more energy.

The current is the same going into the bulb as coming out of it, so electric charge (the lorries) are not 'used up' in a circuit.

We can measure the energy difference between the loaded lorries going into the bulb and the empty ones leaving it using a voltmeter. The voltmeter is connected *across* the bulb to measure how much energy has been transferred to the bulb by comparing the energy (Joules) carried by the lorries (Coulombs) before and after the bulb. *Each Volt represents one Joule transferred by one Coulomb.* The proper name of this is potential difference (because the current has more potential to do work before the bulb than after it) but is often called the voltage.

Questions   Use the lorry model to explain:
1. Why the ammeter readings are the same all the way round a series circuit.
2. Why the total current flowing into a junction is the same as the total current flowing out.
3. Why all the bulbs in a parallel circuit light at full brightness.
4. Why the bulbs get dimmer as you add more in a series circuit.
5. Why the cell goes 'flat' more quickly if you add more bulbs in parallel.

6. Should a 'flat' battery be described as discharged or de-energized? Discuss.
7. This model cannot explain all the features of a circuit. Try to explain:
   a. How the lorries know to save some energy for the next bulb in a series circuit.
   b. Whether it takes time for the first full lorries to reach the bulb and make it light up.
   c. Whether there are full lorries left in the wires when you take the circuit apart.

ELECTRICAL ENERGY   Energy Transfers in Series and Parallel Circuits

1 Volt = 1 Joule of energy per Coulomb of charge

The voltage of a cell is a measure of how many Joules of chemical energy are converted to electrical energy per Coulomb of charge passing though it.

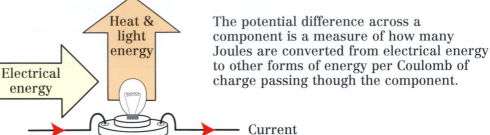

Voltage (sometimes called electromotive force or emf for short) is the energy transferred *to* each Coulomb of charge passing through a source of electrical energy.

Potential difference is the energy given to a device *by* each Coulomb of charge passing through it.

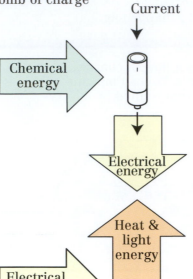

The potential difference across a component is a measure of how many Joules are converted from electrical energy to other forms of energy per Coulomb of charge passing though the component.

A bulb converts electrical energy to thermal and light energy.
A motor converts electrical energy to kinetic energy.
A resistor converts electrical energy to thermal energy.
A loudspeaker converts electrical energy to sound energy.

As energy cannot be created or destroyed all the electrical energy supplied by the cell must be converted into other forms of energy by the other components in the circuit.

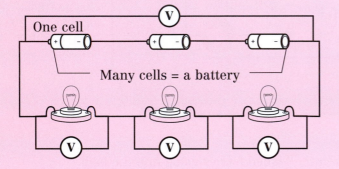

N.B. Voltmeters connected in *parallel*.

One cell

Many cells = a battery

This means that in a *series* circuit the sum of the voltages across the components must equal the voltage across the cell.

> The current is the *same* through all components, the potential difference is *shared* between components.

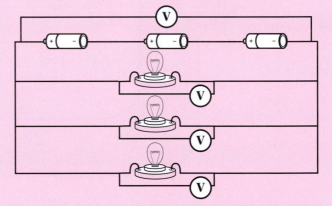

N.B. Voltmeters connected in *parallel*.

In a *parallel* circuit, each Coulomb of charge only passes through one component before returning to the cell. Therefore, it has to give all the energy it carries to that component. Therefore, the potential difference across each component is the same as the potential difference of the cell.

> Potential difference is the *same* across all components, current is *shared* between components.

Questions

1. What is a Joule per Coulomb more commonly called?
2. A cell is labelled 9 V, explain what this means.
3. Explain whether or not voltage splits at a junction in a circuit.
4. A 1.5 V cell is connected in series with a torch bulb. The bulb glows dimly. Explain why adding another cell, in series, will increase the brightness of the bulb.
5. Considering the same bulb as in question 2, adding a second cell in parallel with the first will make no difference to the brightness. Why not?
6. When making electrical measurements we talk about the *current through* a component, but the *voltage across* a component, explain why.
7. Try to write down a formula relating voltage, energy, and charge.

ELECTRICAL ENERGY Resistance

Resistance is a measure of how much energy is needed to make something move or flow.

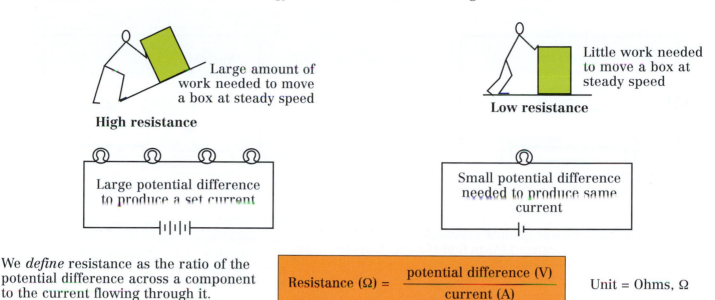

High resistance

Large amount of work needed to move a box at steady speed

Little work needed to move a box at steady speed

Low resistance

Large potential difference to produce a set current

Small potential difference needed to produce same current

We *define* resistance as the ratio of the potential difference across a component to the current flowing through it.

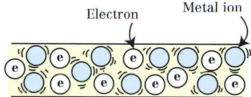

$$\text{Resistance } (\Omega) = \frac{\text{potential difference (V)}}{\text{current (A)}}$$

Unit = Ohms, Ω

i.e. if we have a high resistance then a bigger push is needed to push the current round the circuit. Note that this is not Ohm's Law, just the definition of resistance.

What causes resistance in wires?

In the lorry analogy on p45, the lorry had to use some energy (fuel) to move along the roads (wires). This represents the resistance of the wires.

Electron   Metal ion

Wires have resistance because the electrons moving through the wire bump into the positive metal ions that make up the wire.

The electrons give some of their kinetic energy to the metal ions, which makes them vibrate so electrical energy is converted to thermal energy and the wire gets warm.

The same process happens in a resistor, but the materials are chosen to increase the number of collisions making it more resistant to charge flow.

Factors affecting resistance

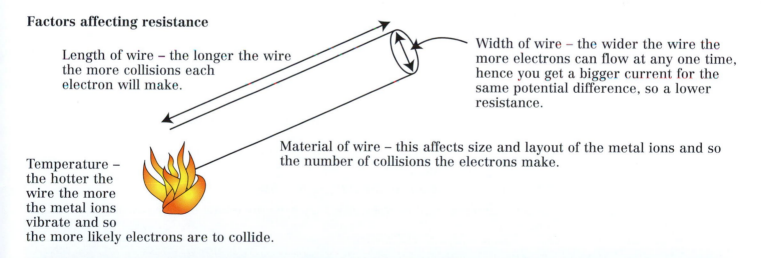

Length of wire – the longer the wire the more collisions each electron will make.

Width of wire – the wider the wire the more electrons can flow at any one time, hence you get a bigger current for the same potential difference, so a lower resistance.

Material of wire – this affects size and layout of the metal ions and so the number of collisions the electrons make.

Temperature – the hotter the wire the more the metal ions vibrate and so the more likely electrons are to collide.

Questions

1. Show that a resistor with 5 V across it and 2 A flowing through it has a resistance of 2.5 Ω.
2. A 12 Ω resistor has 2.4 V across it. Show that the current flowing is 0.2 A.
3. A lamp has a resistance of 2.4 Ω and 5 A flows through it. Show the potential difference is 12 V.
4. The potential difference across the lamp in (3) is doubled. What would you expect to happen to a. the filament temperature, b. the resistance, c. the current?
5. In the following circuit, which resistor has the largest current flowing through it?
6. Why do many electronic devices, e.g. computers, need cooling fans?

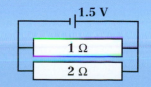

1.5 V

1 Ω

2 Ω

ELECTRICAL ENERGY Electrical Measurements and Ohm's Law

Experimental technique for measuring resistance

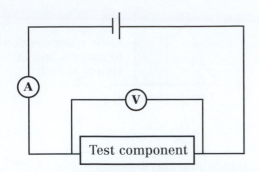

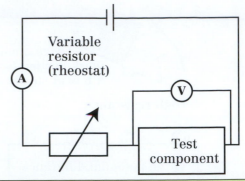

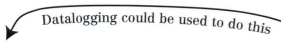

Finding the resistance of a component for a given current

Use meter readings and the formula Resistance (Ω) = potential difference (V) / current (A) to find the resistance.

Datalogging could be used to do this

Use voltmeter and ammeter sensors. Computer software automatically plots voltage vs. current as the variable resistor is altered.

Finding the resistance of a component for a range of currents

① Record voltmeter and ammeter readings for many different settings of the variable resistor in a table.

Voltage across component (Volts)	Current through component (Amps)

② Plot a graph like this

③ Gradient of graph $= \dfrac{\text{change in } y}{\text{change in } x}$

$= \dfrac{\text{change in potential difference, V}}{\text{change in current, A}}$

$=$ Resistance, Ω

The steeper the line the greater the resistance.

N.B. Beware, sometimes these graphs are plotted with the axes reversed. Then the resistance is not the gradient, it is $^{1}/$ gradient.

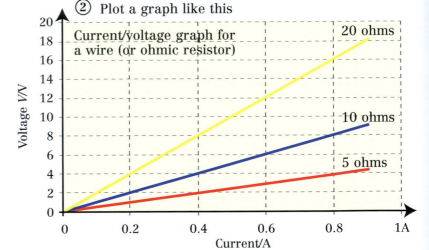

Ohm's Law

A component where the current is directly *proportional* to the voltage is said to obey Ohm's Law and is called *ohmic*.

This means:
1. A graph of *V* vs. *I* is a straight line through the origin.
2. $V = I \times R$ where *R is constant whatever the value of the current or voltage.*

Note that the definition of resistance applies to all components; they are only ohmic if their resistance does not change as the current changes.

Questions
1. Calculate the gradients of the three lines in the graph above and confirm they have the resistances shown.
2. 1.5 A flows in a 1 m length of insulated wire when there is a potential difference of 0.3 V across it.
   a. Show its resistance is 0.2 Ω.
   b. If 0.15 A flows in a reel of this wire when a potential difference of 3 V is placed across it, show that the length of the wire on the reel is 100 m.
3. Current and voltage data is collected from a mystery component using the method above. When plotted the graph looks like this:
   Is the resistance of the component increasing, decreasing, or staying the same as the current increases?

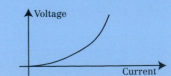

ELECTRICAL ENERGY   Power in (Ohmic) Electrical Circuits

The general definition of power is : $\text{Power (W)} = \dfrac{\text{energy transferred (J)}}{\text{time taken (s)}}$

We also know that 1 Volt = 1 Joule per Coulomb.

As a formula we represent this as $\text{potential difference (Volts)} = \dfrac{\text{energy transferred (Joules)}}{\text{charge passing (Coulombs)}}$

We also know that $\text{current (Amps)} = \dfrac{\text{charge passing (Coulombs)}}{\text{time taken (seconds)}}$

Then $\text{current} \times \text{voltage} = \dfrac{\text{charge passing}}{\text{time taken}} \times \dfrac{\text{energy transferred}}{\text{time taken}} = \dfrac{\text{energy transferred}}{\text{time taken}} = \text{Power}$

Also as voltage = current × resistance then

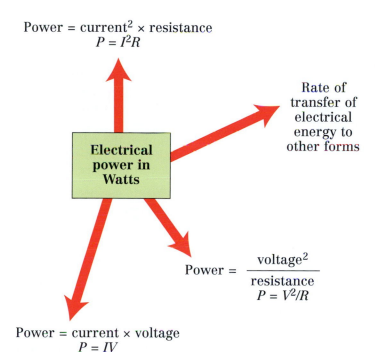

Power = current² × resistance
$P = I^2R$

Electrical power in Watts

Rate of transfer of electrical energy to other forms

$\text{Power} = \dfrac{\text{voltage}^2}{\text{resistance}}$
$P = V^2/R$

Power = current × voltage
$P = IV$

Mains appliances use 230 V and always have a power rating.

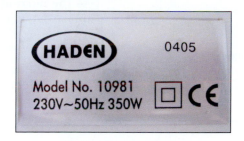

HADEN   0405

Model No. 10981
230V~50Hz 350W CE

We can use this information to calculate the current that flows through them when working normally.

Device	Voltage (V)	Power (W)
Filament bulb	230	60
Energy efficient lamp	230	9
Kettle	230	1500
Microwave oven	230	1600
Electric cooker	230	1000–11 000
TV set	230	30

Questions

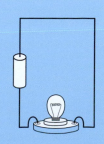

1. Redraw the circuit using standard circuit symbols adding voltmeter to measure the potential difference across the lamp and an ammeter to measure the current through it.
   a. The voltmeter reads 6 V. How many Joules of energy are transferred per Coulomb?
   b. The ammeter reads 2 A. How many Coulombs pass through the lamp each second?
   c. Hence, how many Joules per second are transferred to the lamp?
   d. If the voltmeter now reads 12 V and the ammeter still reads 2 A then how many Joules are transferred to the lamp each second?
2. A 1.5 V cell is used to light a lamp.
   a. How many Joules does the cell supply to each Coulomb of electric charge?
   b. If the current in the lamp is 0.2 A, how many Coulombs pass through it in 5 s?
   c. What is the total energy transferred in this time?
   d. Hence, show the power of the lamp is 0.3 W.
3. A 6 V battery has to light two 6 V lamps fully. Draw a circuit diagram to show how the lamps should be connected across the battery. If each draws a current of 0.4 A when fully lit, explain why the power generated by the battery is 4.8 W.
4. Use the data in the table above to show that the current drawn by an 'energy efficient' lamp is over 6× less than the current drawn by a normal filament bulb.
5. Show that a 60 W lamp with a potential difference of 240 V across it has a resistance of 960 Ω.

ELECTRICAL ENERGY Properties of Some Electrical Components

A graph of voltage against current (or *vice versa*) for a component is called its characteristic.
This circuit can be used to measure the characteristic of component X.

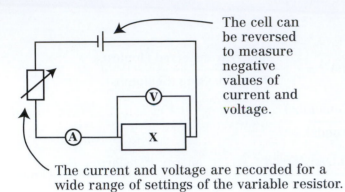

The cell can
be reversed
to measure
negative
values of
current and
voltage.

The current and voltage are recorded for a
wide range of settings of the variable resistor.

N.B. Check carefully whether current or voltage is
plotted on the *x* axis.

If

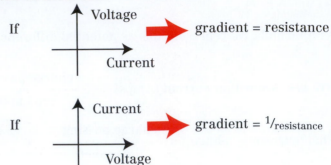

gradient = resistance

If

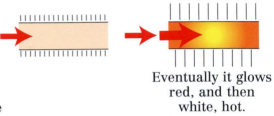

gradient = $^1/_{resistance}$

All the following components are *non-ohmic* as their resistance
is not independent of the current flowing through them.

1. Filament lamp

Symbol

As the current increases, the filament
wire in the bulb becomes hotter.

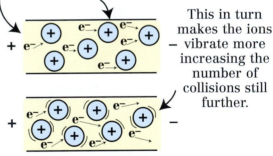

Eventually it glows
red, and then
white, hot.

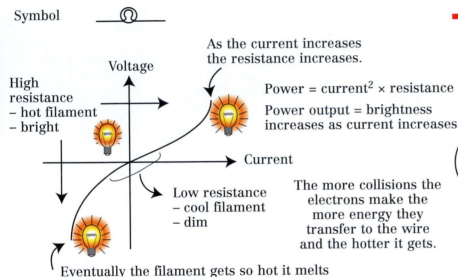

High
resistance
– hot filament
– bright

As the current increases
the resistance increases.

Power = current^2 × resistance

Power output = brightness
increases as current increases

Low resistance
– cool filament
– dim

The more collisions the
electrons make the
more energy they
transfer to the wire
and the hotter it gets.

Eventually the filament gets so hot it melts
and the bulb fails (if the potential difference
exceeds the design potential difference).

Increasing current in the wire means
the electrons make more collisions with
fixed ions in the wire.

This in turn
makes the ions
vibrate more
increasing the
number of
collisions still
further.

Electrons find it harder to move; you need
a greater potential difference to drive the
same current so the resistance increases.

2. Diode

Symbol

A diode only allows current to flow in one direction.

Forward bias – very low resistance, high current flow

Reverse bias – extremely high
resistance, negligible current
flow

In the reverse direction, the
current is almost negligible until
very large voltages are reached
when the diode may fail.

Small
gradient

The forward current 'turns
on' at about 0.5 V and very
large currents can be
achieved by 0.7 V when
there is almost no
resistance in the diode.

Diodes are made from two types of semiconductor

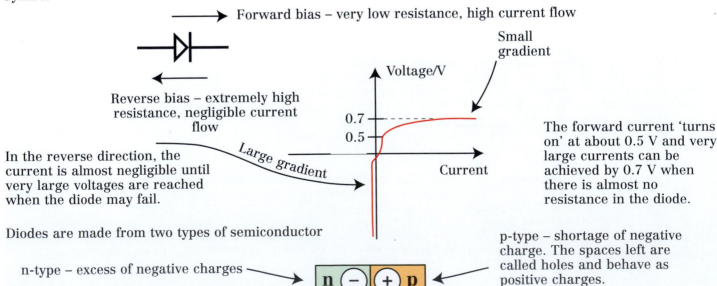

n-type – excess of negative charges

p-type – shortage of negative
charge. The spaces left are
called holes and behave as
positive charges.

Electrons will flow n → p and holes p → n. Therefore, the diode will conduct when the n-type end is negative
and the p-type end is positive.

3. Light dependent resistor (LDR)

Symbol

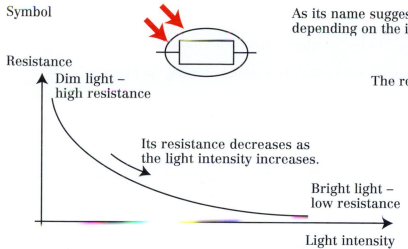

As its name suggests this is a resistor whose resistance changes depending on the intensity of the light falling on it.

Resistance

Dim light – high resistance

Its resistance decreases as the light intensity increases.

Bright light – low resistance

Light intensity

The resistance changes in a non-uniform way.

Used to control electrical circuits that need to respond to varying light levels, e.g. switching on lights automatically at night.

4. Thermistor

Symbol

Resistance

This is a resistor whose resistance varies depending on temperature. Its resistance decreases as temperature increases.

When the thermistor is cold, it has a high resistance

The resistance changes in a non-uniform way

When the thermistor is hot, it has a low resistance

Temperature

Notice that this is the opposite behaviour to a wire, whose resistance increases with increasing temperature.

Thermistors are used to control circuits that need to respond to temperature changes, e.g. to switch off a kettle.

Questions
1. Draw the circuit symbols for: a. A filament lamp. b. An LDR. c. A thermistor. d. A diode.
2. Sketch a graph of current against voltage for a filament lamp. Explain in terms of the motion of electrons through the filament the shape of the graph.
3. Show that a thermistor with a potential difference of 3 V across it and a current of 0.2 A flowing through it has a resistance of 15 Ω. If the temperature of the thermistor was raised, what would you expect to happen to its resistance?
4. Show that an LDR with a potential difference of 1.5 V across it and a current of 7.5×10^{-3} A (7.5 mA) flowing through it has a resistance of 200 Ω. If the LDR is illuminated with a brighter light, with the same potential difference across it what would you expect to happen to the current flowing in it and why?
5. Sketch a graph of current against voltage (both positive and negative values) for a diode. Use it to explain why a diode only passes current in one direction.
6. Consider the following circuits. In which circuit will the ammeter show the greatest current?
7. A student plans to use a thermistor to investigate how the temperature of the water in a kettle varies with time after it is switched on.
   a. Draw a circuit involving an ammeter and voltmeter the student could use.
   b. Explain how they would use the ammeter and voltmeter readings together with a graph like the one printed above on this page, to find the temperature of the water at any given time.

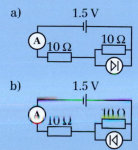

a) 1.5 V
$10\,\Omega$ $10\,\Omega$

b) 1.5 V
$10\,\Omega$ $10\,\Omega$

ELECTRICAL ENERGY   Potential Dividers

Two resistors in series form a potential divider.

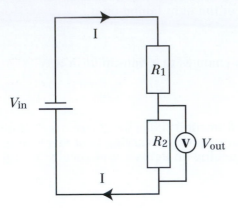

The potential difference of the cell, V_{in}, is divided between the two resistances in the ratio of their resistances.

The output voltage can be calculated using the formula:

$$V_{out} = V_{in} \times \left(\frac{R_2}{R_1 + R_2} \right)$$

V_2 (output potential difference) across here

A rheostat can be used as a potential divider.

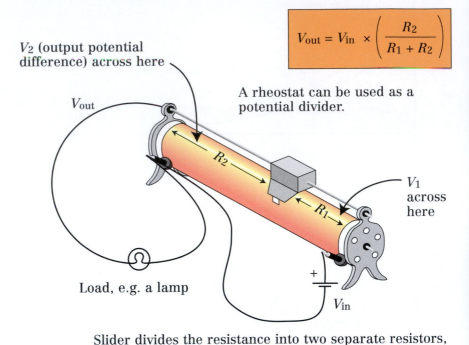

V_1 across here

The current in both resistors is the same as they are in series. The resistor with the greater value will take more voltage to drive the current through it so has the greater potential difference across it.

Load, e.g. a lamp

If one of the resistors is a variable resistor, the ratio of the resistances can be altered. This means you can have a variable output voltage.

Slider divides the resistance into two separate resistors, R_1 and R_2. The position of the slider determines the relative size, hence ratio, of the two resistances.

If one of the resistors is an LDR, the output voltage depends on the light level.
LDR has a large resistance in the dark.

If one of the resistors is a thermistor, the output voltage depends on temperature.

Thermistors have a high resistance in the cold.

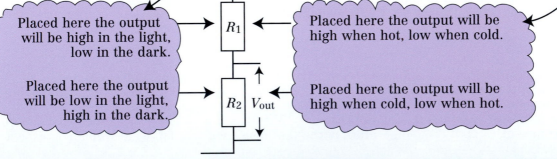

Placed here the output will be high in the light, low in the dark.

Placed here the output will be high when hot, low when cold.

Placed here the output will be low in the light, high in the dark.

Placed here the output will be high when cold, low when hot.

Questions
1. Use the formula above to calculate V_{out} if R_1 and R_2 in the circuit provided have the following values:
   a. R1 = 10 Ω, R2 = 20 Ω.   b. R1 = 20 Ω, R2 = 10 Ω.   c. R1 = 1 kΩ, R2 = 5 kΩ.
   d. R1 = 1.2 kΩ, R2 = 300 Ω.
2. For each of the pairs of resistors in question 1, decide whether R1 or R2 has the greater potential difference across it.
3. An LDR has a resistance of 1000 Ω in the light and 100 000 Ω in the dark. In the circuit, the variable resistor is set to 5000 Ω. Calculate V_{out} in the light and in the dark. If the resistance of the variable resistor is reduced, will the values of V_{out} increase or decrease?
4. Draw a potential divider circuit where the output rises as the temperature rises. Suggest a practical application of this circuit.

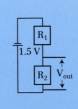

ELECTRICAL ENERGY Electric Cells, Alternating and Direct Current

Electron flow allows chemicals inside the cell to react.

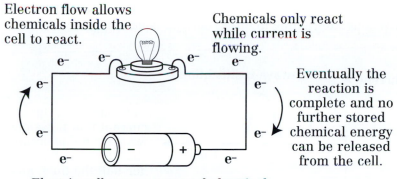

Electric cells convert stored chemical energy into electrical energy.

Chemicals only react while current is flowing.

Eventually the reaction is complete and no further stored chemical energy can be released from the cell.

Circuit symbols

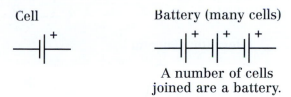

| Cell | Battery (many cells) |

A number of cells joined are a battery.

Current produced by cells only ever flows one-way.
Conventional current – positive to negative

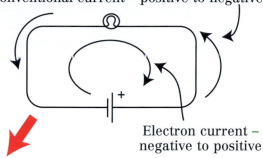

Electron current – negative to positive

The capacity of a cell is measured in *Amp-hours*. 1 Amp-hour means the cell can deliver a current of 1 Amp for 1 hour. Since 1 Amp-hour = 3600 C the energy stored (Joules) in a cell can be calculated from voltage (V) × capacity (Amphours) × 3600 C.

'Rechargeable' cells use another source of electricity, often the mains, to force the electrons the 'wrong way' around the circuit. This, in a specially designed cell, reverses the chemical reaction, storing the electrical energy as chemical energy. It would be more correct to say the battery has been 're-energized'.

Both these currents are *direct* currents (or d.c.). They flow consistently in one direction.

Compare this with mains electricity.

This is *alternating current or a.c.*

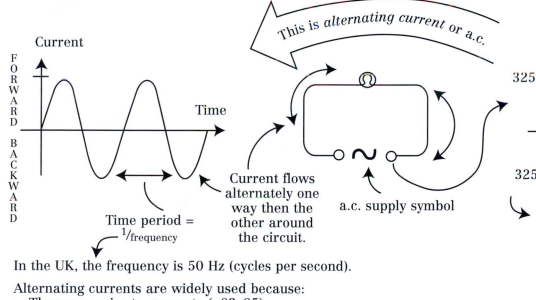

Current flows alternately one way then the other around the circuit.

a.c. supply symbol

Time period = $^1/_{\text{frequency}}$

In the UK, the frequency is 50 Hz (cycles per second).

Alternating currents are widely used because:
• They are easier to generate (p83–85).
• They are easier to distribute (p86–7) than direct currents.
Many devices need direct current to work so alternating current often has to be converted to direct current (p54).

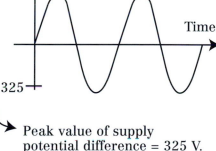

P.d. across supply

325

325

Peak value of supply potential difference = 325 V. But, because it only has the peak value for a short time the supply transfers an equivalent power to a 230 V direct current supply.

(p83–85), (p86–7), (p54).

Questions
1. Sketch a labelled graph of the variation of supply potential difference with time (for 10 seconds) for alternating current of frequency 2 Hz, and peak value 3 V. Add to the graph a line showing the output from a battery of terminal potential difference 2 V.
2. The capacities of two cells are AA = 1.2 Amp-hours and D = 1.4 Amp-hours. How long will each cell last when supplying:
   a. A current of 0.5 A to a torch bulb?   b. 50 mA to a light emitting diode?
3. Some people claim that battery powered cars do not cause any pollution. A battery is just a store of electrical energy so where do battery-powered cars really get their energy from? Hence, are they really non-polluting, or is the pollution just moved elsewhere?
4. Draw up a table of advantages and disadvantages of batteries compared to mains electricity. Consider relative cost, how they are used, potential power output, and impact on the environment.

ELECTRICAL ENERGY   Diodes, Rectification and Capacitors

Although alternating current is easier to generate and distribute, many appliances, especially those with microchips, need direct current. The process of converting alternating to direct current is called *rectification*.

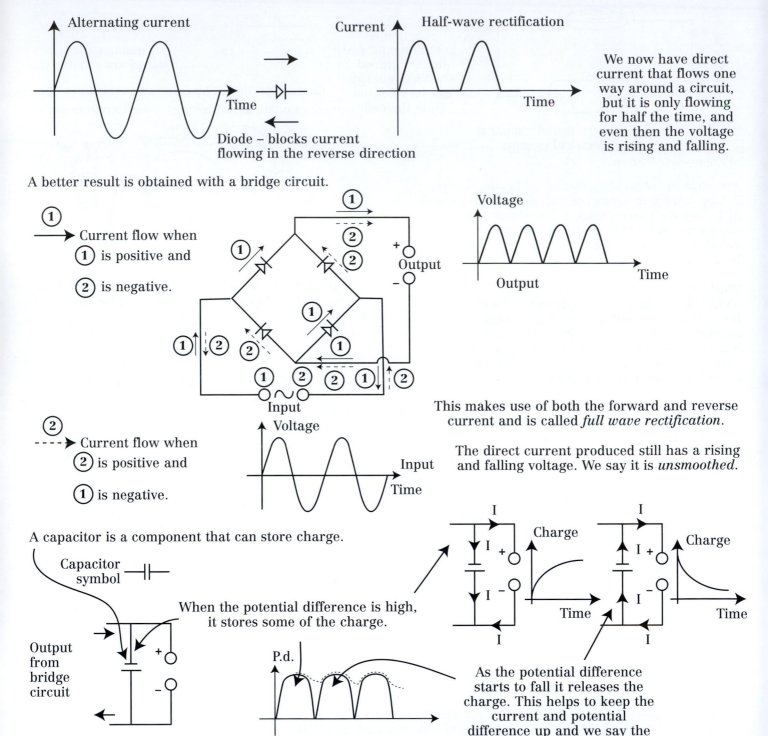

Alternating current

Time

Diode – blocks current flowing in the reverse direction

Current   Half-wave rectification

Time

We now have direct current that flows one way around a circuit, but it is only flowing for half the time, and even then the voltage is rising and falling.

A better result is obtained with a bridge circuit.

① Current flow when ① is positive and ② is negative.

Voltage

Output

Time

Voltage

Input

Time

Input

② Current flow when ② is positive and ① is negative.

This makes use of both the forward and reverse current and is called *full wave rectification*.

The direct current produced still has a rising and falling voltage. We say it is *unsmoothed*.

A capacitor is a component that can store charge.

Capacitor symbol

Output from bridge circuit

When the potential difference is high, it stores some of the charge.

P.d.

As the potential difference starts to fall it releases the charge. This helps to keep the current and potential difference up and we say the direct current is smoothed.

I   Charge

Time

I   Charge

Time

Questions
1. What is a diode?
   a. Complete the graphs in the circuit below to show the effect of the diode.
   b. Why is the output an example of direct current? Why do we say it is 'unsmoothed'?
   c. If the diode were reversed what would be the effect, if any, on the direct current output?
2. What name do we give a device that stores charge?
3. Explain the difference between full wave rectification and half-wave rectification. Illustrate your answer with voltage–time graphs.
4. Draw a circuit that produces full wave rectification. Show how the current flows through the circuit.

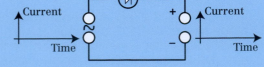

Current

Time

Current

Time

ELECTRICAL ENERGY   Mains Electricity and Wiring

⚠ N.B. Never inspect any part of mains wiring without first switching off at the main switch next to the electricity meter.

The UK mains electricity supply is alternating current varying between +325 V and –325 V with a frequency of 50 Hz. It behaves as the equivalent of 230 V direct current.

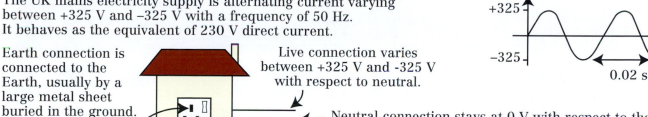

Voltage
+325
Time/s
–325
0.02 s

Earth connection is connected to the Earth, usually by a large metal sheet buried in the ground.

Live connection varies between +325 V and -325 V with respect to neutral.

Neutral connection stays at 0 V with respect to the Earth.

The Earth acts as a vast reservoir of charge; electrons can flow into it or out of it easily. Therefore, its potential is 0 V.

Touching the live wire is dangerous because if you are also connected to Earth, electrons can flow across the potential difference between Earth and live, through you. This will give you a shock.

When the live is at +325 V electrons flow from neutral to live.

Live

When the live is at –325 V electrons flow from live to neutral.

Neutral

Plug sockets have three terminals:

Therefore, switches and fuses are always placed in the live wire.

Yellow/green wire – Earth wire. This can never become live, as it will always conduct electrons into or out of the Earth, so its potential will always be 0 V. All metal cased appliances must have an Earth connection.

Ensure wires are connected firmly, with no stray metal conductors.

Three pin plug

Ensure cable grip is tightened, so if the cable is pulled, the conductors are not pulled out of their sockets.

Most appliances use three core cables.

Brown wire, live.

Blue wire, neutral. Needed to complete the circuit.

Inner insulation to separate the conductors.

Outer insulation to protect the conductors from damage.

The current drawn by an appliance can be calculated using the equation:

$$\text{Current} = \frac{\text{Power}}{\text{voltage}} = \frac{\text{Power}}{230 \text{ V}}$$

Power ratings can be found on the information label on the appliance.

Questions
1. What colours are the following electrical wires: live, neutral, Earth?
2. My kettle has a power output of 1 kW and my electric cooker 10 kW. What current will each draw? Why does the cooker need especially thick connecting cables?
3. Some countries use 110 V rather than 230 V for their mains supply. Suggest how the thickness of the conductors in their wiring would compare to the conductors used in the UK. How will this affect the cost of wiring a building? What advantages does using a lower voltage have?
4. Study this picture of a three-pin plug – how many faults can you find?
5. Placing a light switch in the neutral wire will not affect the operation of the light but could make changing a bulb hazardous. Why?

ELECTRICAL ENERGY Electrical Safety

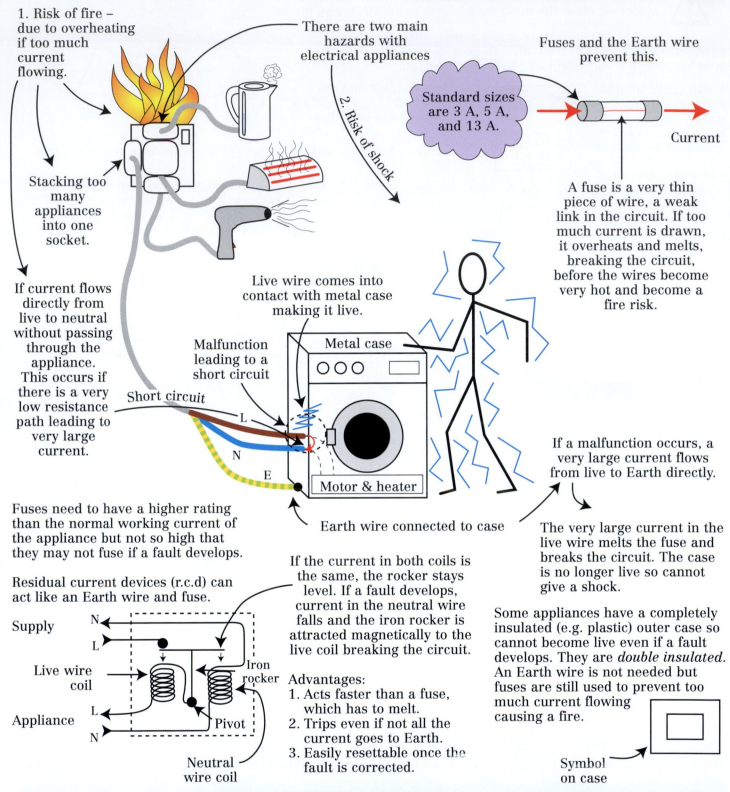

1. Risk of fire – due to overheating if too much current flowing.

There are two main hazards with electrical appliances

Fuses and the Earth wire prevent this.

Standard sizes are 3 A, 5 A, and 13 A.

Current

Stacking too many appliances into one socket.

2. Risk of shock

A fuse is a very thin piece of wire, a weak link in the circuit. If too much current is drawn, it overheats and melts, breaking the circuit, before the wires become very hot and become a fire risk.

If current flows directly from live to neutral without passing through the appliance. This occurs if there is a very low resistance path leading to very large current.

Live wire comes into contact with metal case making it live.

Malfunction leading to a short circuit

Short circuit

Metal case

L

N

E

Motor & heater

Earth wire connected to case

If a malfunction occurs, a very large current flows from live to Earth directly.

The very large current in the live wire melts the fuse and breaks the circuit. The case is no longer live so cannot give a shock.

Fuses need to have a higher rating than the normal working current of the appliance but not so high that they may not fuse if a fault develops.

Residual current devices (r.c.d) can act like an Earth wire and fuse.

Supply

N
L

Live wire coil

L

Appliance

N

Iron rocker

Pivot

Neutral wire coil

If the current in both coils is the same, the rocker stays level. If a fault develops, current in the neutral wire falls and the iron rocker is attracted magnetically to the live coil breaking the circuit.

Advantages:
1. Acts faster than a fuse, which has to melt.
2. Trips even if not all the current goes to Earth.
3. Easily resettable once the fault is corrected.

Some appliances have a completely insulated (e.g. plastic) outer case so cannot become live even if a fault develops. They are *double insulated*. An Earth wire is not needed but fuses are still used to prevent too much current flowing causing a fire.

Symbol on case

Questions
1. Choose (from 3 A, 5 A, and 13 A) the most appropriate fuse for the following:
   a. An electric iron of power output 800 W.
   b. A table lamp of power output 40 W.
   c. A washing machine of total power 2500 W.
2. Explain why a fuse must always be placed in the live wire.
3. Explain why a double insulated appliance does not need an Earth wire, but does need a fuse.
4. The maximum current that can be safely drawn from a normal domestic socket is 13 A. At my friend's house, I notice a 2.5 kW electric fire, an 800 W iron, and three 100 W spot lamps all connected to a single socket. What advice should I give my friend? Use a calculation to support your answer.
5. While using my electric lawnmower I cut the flex, and the live wire comes into contact with the damp grass. An r.c.d will make this wire safe very quickly. What is an r.c.d and how does it work in this case?

ELECTRICAL ENERGY Electron Beams

Not all electric currents flow in wires. It is possible to produce a beam of electrons travelling through a vacuum.

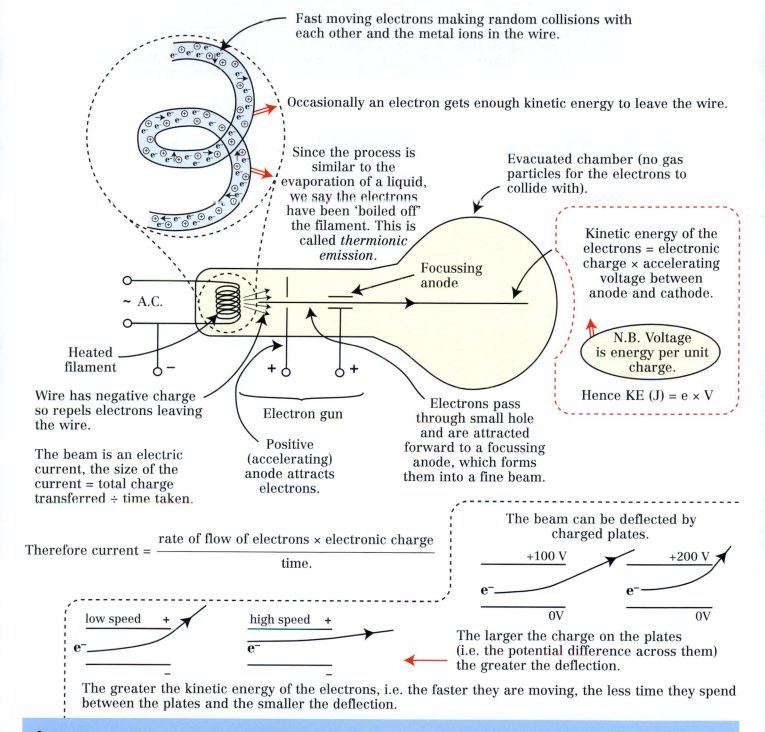

Fast moving electrons making random collisions with each other and the metal ions in the wire.

Occasionally an electron gets enough kinetic energy to leave the wire.

Since the process is similar to the evaporation of a liquid, we say the electrons have been 'boiled off' the filament. This is called *thermionic emission*.

Evacuated chamber (no gas particles for the electrons to collide with).

Kinetic energy of the electrons = electronic charge × accelerating voltage between anode and cathode.

N.B. Voltage is energy per unit charge.

Hence KE (J) = e × V

Focussing anode

~ A.C.

Heated filament

Wire has negative charge so repels electrons leaving the wire.

Electron gun

Positive (accelerating) anode attracts electrons.

Electrons pass through small hole and are attracted forward to a focussing anode, which forms them into a fine beam.

The beam is an electric current, the size of the current = total charge transferred ÷ time taken.

Therefore current = $\dfrac{\text{rate of flow of electrons} \times \text{electronic charge}}{\text{time.}}$

The beam can be deflected by charged plates.

+100 V

e⁻

0V

+200 V

e⁻

0V

The larger the charge on the plates (i.e. the potential difference across them) the greater the deflection.

low speed   +

e⁻

—

high speed   +

e⁻

—

The greater the kinetic energy of the electrons, i.e. the faster they are moving, the less time they spend between the plates and the smaller the deflection.

Questions

1. Describe the process of thermionic emission. Why is it important that the electron beam be produced in a vacuum?
2. What would happen to the kinetic energy of the electrons produced by an electron gun if the potential difference between the heated filament and the accelerating anode was increased?
3. What would happen to the charge transferred per second (the current) in the electron beam if the heater temperature was increased but the accelerating potential was not changed? Would the kinetic energy of the electrons change?
4. Given that the charge of one electron is 1.6 × 10⁻¹⁹ C, show that the kinetic energy of an electron in the beam is 3.2 × 10⁻¹⁷ J when the accelerating potential is 200 V.

5. If the current in the electron beam is 2 mA, show that the number of electrons boiled off the filament each second is about 1.3 × 10¹⁶ [charge on the electron = 1.6 × 10⁻¹⁹ C].
6. Use the answers to questions 4 and 5 to show that the total energy delivered by the beam per second (i.e. its power) is 0.4 W.
7. An electron beam passes through two charged plates as shown in the diagram. What would be the effect on the deflection of:
   a. Increasing the potential difference across the deflecting plates?
   b. Decreasing the accelerating voltage across the electron gun?

MAGNETIC FIELDS Magnetism and the Earth's Magnetic Field

A magnetic field is a region of space in which magnets and magnetic materials feel forces. The only magnetic materials are iron, steel, nickel, and cobalt. We represent magnetic fields by drawing magnetic field lines.

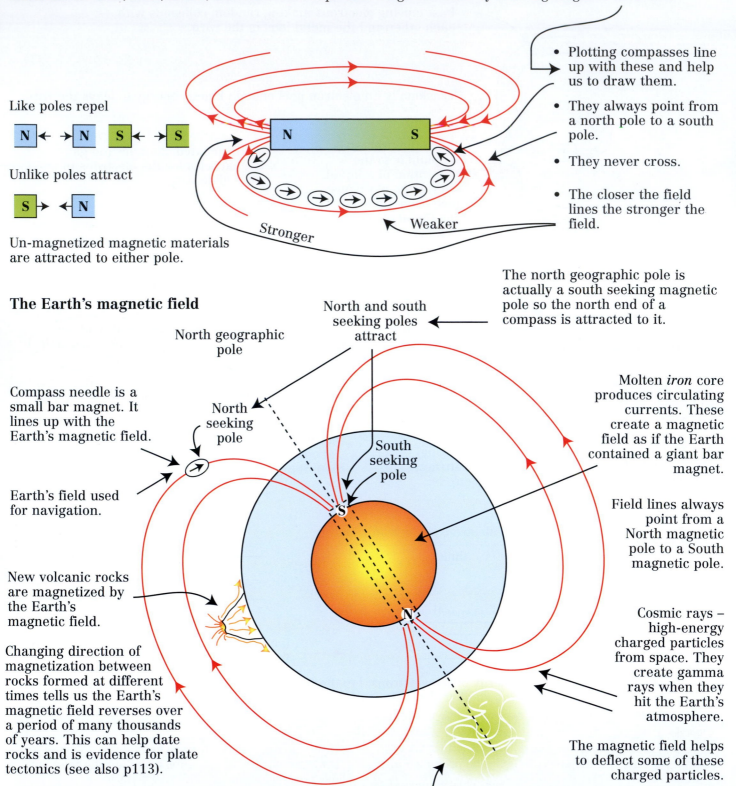

Like poles repel

N ← → N   S ← → S

Unlike poles attract

S → ← N

Un-magnetized magnetic materials are attracted to either pole.

N   S

Stronger   Weaker

- Plotting compasses line up with these and help us to draw them.
- They always point from a north pole to a south pole.
- They never cross.
- The closer the field lines the stronger the field.

The Earth's magnetic field

North geographic pole

North and south seeking poles attract

North seeking pole

South seeking pole

Compass needle is a small bar magnet. It lines up with the Earth's magnetic field.

Earth's field used for navigation.

New volcanic rocks are magnetized by the Earth's magnetic field.

Changing direction of magnetization between rocks formed at different times tells us the Earth's magnetic field reverses over a period of many thousands of years. This can help date rocks and is evidence for plate tectonics (see also p113).

The north geographic pole is actually a south seeking magnetic pole so the north end of a compass is attracted to it.

Molten *iron* core produces circulating currents. These create a magnetic field as if the Earth contained a giant bar magnet.

Field lines always point from a North magnetic pole to a South magnetic pole.

Cosmic rays – high-energy charged particles from space. They create gamma rays when they hit the Earth's atmosphere.

The magnetic field helps to deflect some of these charged particles.

The Earth's magnetic field interacts with charged particles from the Sun. They are channelled to the poles where they interact with molecules in the atmosphere making them glow. This is the aurora.

Questions
1. What is a magnetic field? Make a list of three properties of magnetic field lines.
2. Make a list of the four magnetic materials. How could you test an unknown material to discover whether it is one of the four in the list?
3. Using a magnet how would you tell if a piece of steel was magnetized or un-magnetized?
4. If the Earth's magnetic field were to disappear, it would be very bad news for our health. Explain why. (You might need to look at p69.)
5. Why might a magnetic compass not work very well close to the North or South Pole?

MAGNETIC FIELDS Electromagnetism and The Motor Effect

A current carrying wire produces a magnetic field around it.

Plotting compasses line up with the field.

Wire

Circular field lines (getting further apart as the field gets weaker further from the wire).

Current

A coil (or solenoid) produces a magnetic field through and around it.

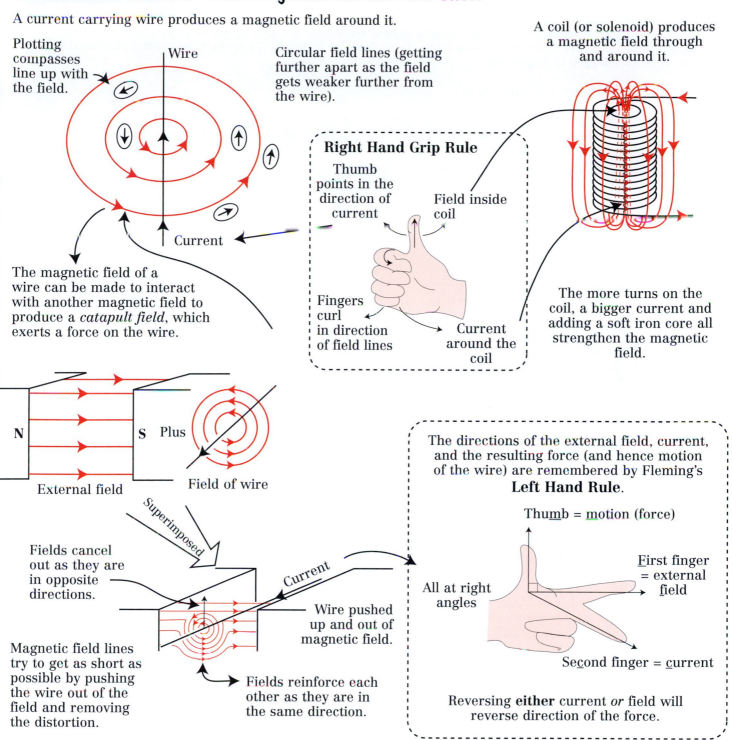

The magnetic field of a wire can be made to interact with another magnetic field to produce a *catapult field*, which exerts a force on the wire.

Right Hand Grip Rule

Thumb points in the direction of current

Field inside coil

Fingers curl in direction of field lines

Current around the coil

The more turns on the coil, a bigger current and adding a soft iron core all strengthen the magnetic field.

N   S   Plus

External field

Field of wire

Superimposed

Current

Fields cancel out as they are in opposite directions.

Magnetic field lines try to get as short as possible by pushing the wire out of the field and removing the distortion.

Wire pushed up and out of magnetic field.

Fields reinforce each other as they are in the same direction.

The directions of the external field, current, and the resulting force (and hence motion of the wire) are remembered by Fleming's **Left Hand Rule**.

Thumb = motion (force)

First finger = external field

All at right angles

Second finger = current

Reversing **either** current *or* field will reverse direction of the force.

If the current is parallel to the external magnetic field the two magnetic fields are at right angles to each other and cannot interact so no force is produced.

Size of the force can be increased by:
- Using a larger current
- Using a stronger external field

Questions
1. In what ways are the fields around a bar magnet and around a long coil (solenoid) similar and in what ways are they different?
2. What would happen to the direction of the magnetic field lines around a wire, or through a coil, if the current direction reverses?
3. Make a list of five uses for an electromagnet and suggest why electromagnets are often more useful than permanent magnets.
4. What happens to the direction of the force on a current carrying wire if both the field and current directions are reversed?
5. Copy the diagrams (right) and add an arrow to show the direction of the force on the wire.

a)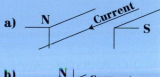

b)

THERMAL ENERGY   Heat and Temperature – What is the Difference?

All energy ultimately ends up as heat. In most energy transfers, a proportion ends up as heat energy, and often this is not useful. Sometimes we want to encourage heat transfers, in cooking for example, and sometimes discourage them, in preventing heat losses from your home for example. Therefore, understanding heat energy and how it is transferred is important.

Are heat and temperature the same thing?

> We define heat energy as the total kinetic energy of the particles in a substance (in Joules).

Identical kettles, both switched on for the same time.

Same electrical energy supplied to both. Therefore, the water gains the same amount of heat energy. Therefore, the total kinetic energy of the particles in both kettles will be the same.

Particles move faster in this kettle – large average kinetic energy.

Particles move more slowly in this kettle – lower average kinetic energy.

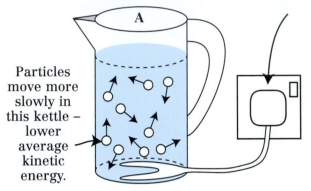

This kettle has half the number of particles so to make a fair comparison we measure the average energy per particle.

We have a special name for this average; we call it *temperature*.

Experience tells us that kettle B is hotter than A. This means that the particles in B have a higher average kinetic energy than those in A. This is reasonable because the same amount of energy is spread over fewer particles in B than in A.

Temperature differences tell us how easily heat is transferred. The bigger the temperature *difference* between an object and its surroundings the more easily heat will be transferred.

Heat always flows from hot to cold

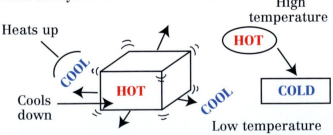

> Temperature (in Kelvin) is proportional to the average kinetic energy of the particles.

Temperature Scales

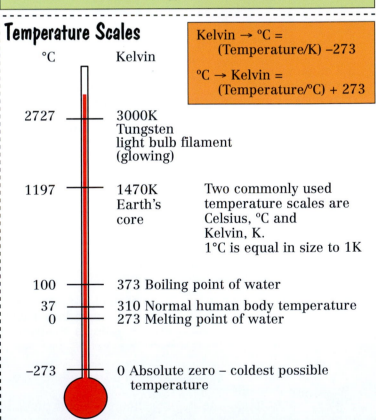

Kelvin → °C = (Temperature/K) –273

°C → Kelvin = (Temperature/°C) + 273

°C	Kelvin	
2727	3000K	Tungsten light bulb filament (glowing)
1197	1470K	Earth's core
100	373	Boiling point of water
37	310	Normal human body temperature
0	273	Melting point of water
–273	0	Absolute zero – coldest possible temperature

Two commonly used temperature scales are Celsius, °C and Kelvin, K.
1°C is equal in size to 1K

Questions
1. Explain how a bath of water at 37°C can have more heat energy than an electric iron at 150°C.
2. A red-hot poker placed in a small beaker of water will make the water boil, but placed in a large bucket of water the temperature of the water only rises a few degrees, why?
3. Which should lose heat faster, a mug of tea at 80°C in a fridge at 5°C, or the same mug of tea at 40°C, placed in a freezer at –10°C?

Questions
1. Convert the following into Kelvin: 42°C, 101°C, –78°C, –259°C.
2. Convert the following into °C: 373K, 670K, 54K, 4K.

THERMAL ENERGY Specific and Latent Heat

When an object cools, it transfers heat to its surroundings.

Consider

1 kg 99°C
1 kg 1°C
2 kg 50°C

The total heat energy available has been shared among all the particles.

However

1 kg aluminium at 99°C

1 kg water at 1°C

Temperature of water / aluminium when they come to equilibrium <50°C (and no heat has been lost from the container).

This tells us that 1 kg of aluminium has less heat energy stored in it than 1 kg of water, so the average kinetic energy (temperature) of the particles when mixed is less. We say aluminium has a lower specific heat capacity than water.

Since temperature is proportional to the average kinetic energy of the particles we are actually measuring the energy needed to increase the average kinetic energy of the particles by a set amount. This will depend on the structure of the material, i.e. what it is made of and whether it is a solid, liquid, or gas. Therefore, *all materials have their own specific heat capacities.*

Specific heat capacity is a measure of how much heat energy 1 kg of a material can hold, defined as:

The energy needed to be supplied to raise the temperature of 1 kg of a material by 1K.

Units J/kgK

Energy supplied (J) = mass (kg) × specific heat capacity (J/kgK) × temperature change (K).

$$\Delta E = m \times shc \times \Delta T$$

Latent heat is a measure of the energy needed to completely melt or boil 1 kg of a material.

Energy (J) = mass (kg) × specific latent heat (J/kg)

$$\Delta E = m \times slh$$

Units J/kg

Specific latent heat depends on the strength and number of intermolecular bonds between molecules, so depends on the material and its state.

Heat energy used to raise average kinetic energy of molecules, therefore temperature rises

Energy transferred is used to break bonds between molecules, not to increase their kinetic energy (temperature).

Temperature does not rise above 100°C until all the water is evaporated.

E.g. for water

Temperature/°C

Temperature does not rise above 0°C until all the ice is melted.

150

100

0

−10

MELTING

LIQUID

BOILING

GAS

SOLID

a   b   c   d

See questions below

Energy supplied/J

Questions
1. What happens to the average kinetic energy of the particles in material when the temperature rises?
2. A pan of boiling water stays at 100°C until all the water has evaporated. Why?
3. Explain why adding ice to a drink cools it down.
4. Given that specific heat capacity of water = 4200 J/kgK and of steam = 1400 J/kgK and that the specific latent heat for melting ice is 334 000 J/kg Iand for boiling water = 2 260 000 J/kg show that if the graph in the text above represents 2.5 kg of water:
   a. The energy supplied between a and b is 835 000 J.
   b. The energy supplied between b and c is 1 050 000 J.
   c. The energy supplied between c and d is 5 650 000 J.
5. A student finds that it takes 31 500 J to heat a 1.5 kg block of aluminium from 21°C to 44°C. Show that the specific heat capacity of aluminium is about 900 J/kgK.

THERMAL ENERGY Heat Transfer 1 – Conduction

Conduction is the transfer of thermal energy from a high temperature region to low temperature region by the transfer of kinetic energy between particles in a material.

High temperature – high average kinetic energy

 KE transferred by collisions

Low temperature – low average kinetic energy

This is especially effective in solids where the particles are firmly bonded together.

Metals and graphite have free electrons, which can move and carry heat energy. They are very good conductors.

Good conductors transfer heat energy well.

 Cold metal — Feels very cold

 Hot metal — Feels very hot

Insulators do not transfer heat well.

 Cold wood — Feels cool

Hot wood

 Feels warm

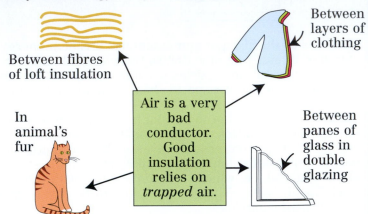

Between fibres of loft insulation

In animal's fur

Between layers of clothing

Air is a very bad conductor. Good insulation relies on *trapped* air.

Between panes of glass in double glazing

Heat Transfer 2 – Convection

Convection happens in *fluids* because the particles can move and carry heat energy away from the source. Hot fluids have a lower density than cool fluids.

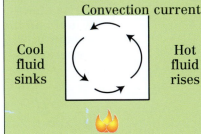

Convection current

Cool fluid sinks

Hot fluid rises

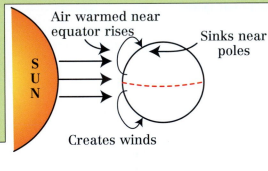

Air warmed near equator rises

Sinks near poles

SUN

Creates winds

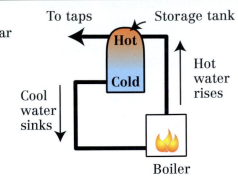

To taps   Storage tank

Hot

Cold

Hot water rises

Cool water sinks

Boiler

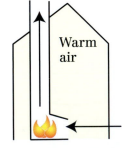

Warm air

Fire burns better as fresh air is drawn through it.

Cool air

Examples of convection

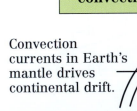

Convection currents in Earth's mantle drives continental drift.

Crust

Core Hot

Mantle

Questions – conduction

1. Using the idea of particles explain why metals are such good conductors of heat and why air is a bad conductor.
2. Air is often trapped, for example between layers of clothing, to reduce conduction. Make a list of five places where air is trapped to prevent conduction.
3. Stuntmen can walk (quickly) across a bed of burning coals without injury, yet briefly touching a hot iron causes a painful burn, why?

Questions – convection

1. Explain the result opposite using the ideas of conduction and convection:

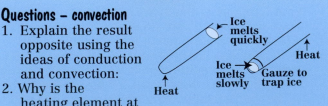

Ice melts quickly

Heat

Ice melts slowly

Gauze to trap ice

Heat

2. Why is the heating element at the bottom of a kettle?
3. Suggest what causes currents in the oceans (in detail).

THERMAL ENERGY   Heat Transfer 3 – Radiation

Thermal radiation is the transfer of heat energy by (infrared) electromagnetic waves (see p30 and 32).

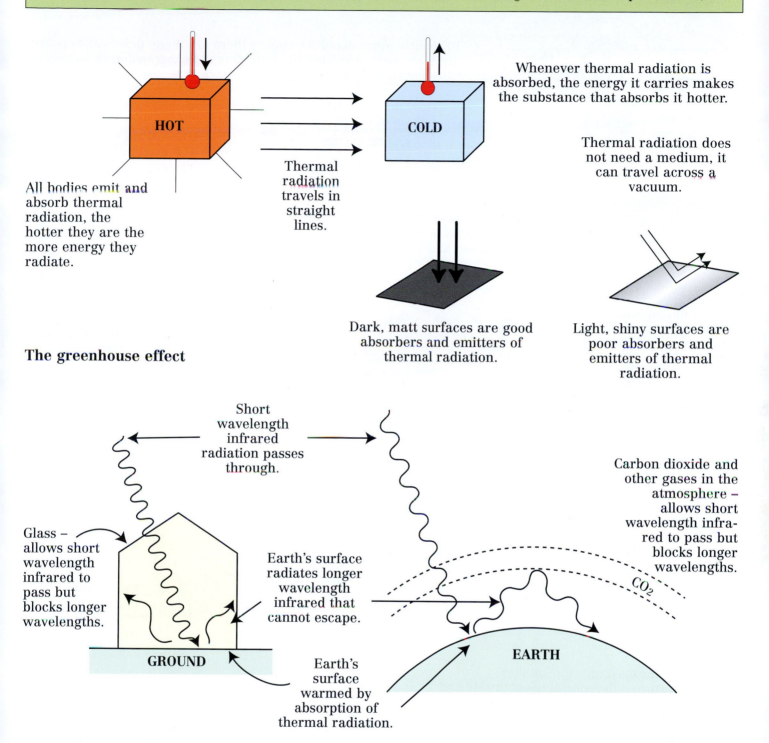

All bodies emit and absorb thermal radiation, the hotter they are the more energy they radiate.

Thermal radiation travels in straight lines.

Whenever thermal radiation is absorbed, the energy it carries makes the substance that absorbs it hotter.

Thermal radiation does not need a medium, it can travel across a vacuum.

Dark, matt surfaces are good absorbers and emitters of thermal radiation.

Light, shiny surfaces are poor absorbers and emitters of thermal radiation.

The greenhouse effect

Short wavelength infrared radiation passes through.

Glass – allows short wavelength infrared to pass but blocks longer wavelengths.

Earth's surface radiates longer wavelength infrared that cannot escape.

GROUND

Earth's surface warmed by absorption of thermal radiation.

Carbon dioxide and other gases in the atmosphere – allows short wavelength infra-red to pass but blocks longer wavelengths.

CO_2

EARTH

Temperature rises due to the reabsorption of the trapped heat radiation. This leads to global warming.

Questions
1. By which method of heat transfer does the heat from Sun reach Earth? How can you tell?
2. Why are solar panels fixed to roofs and designed to heat water painted black?
3. Why are many teapots made of shiny steel?
4. Explain why it is important to reduce the amount of carbon dioxide we pump into the atmosphere.
5. Look at the following diagram of a thermos flask and explain why:
   a. There is a vacuum between the walls of the flask.
   b. The walls of the flask are shiny.
   c. The drink stays hotter longer if the stopper is put in.
   d. Liquid nitrogen (boiling point 77K, –196°C) stays as a liquid in the flask for a long time, but rapidly boils and evaporates if poured out.

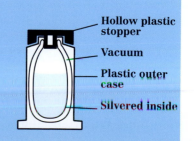

Hollow plastic stopper

Vacuum

Plastic outer case

Silvered inside

THERMAL ENERGY Reducing Energy Wastage in Our Homes

Reducing our demand for energy is as important in reducing greenhouse gas emissions as finding renewable energy resources. Although energy is conserved, we often convert it to forms that are not useful. For example:

Useful light

Heat – not useful

100% electrical

To reduce our demand for energy there are many practical steps we can take, many involving the reduction of heat transfer. However, householders must also consider payback time.

$$\frac{\text{Cost of installing}}{\text{annual saving}} = \text{payback time in years.}$$

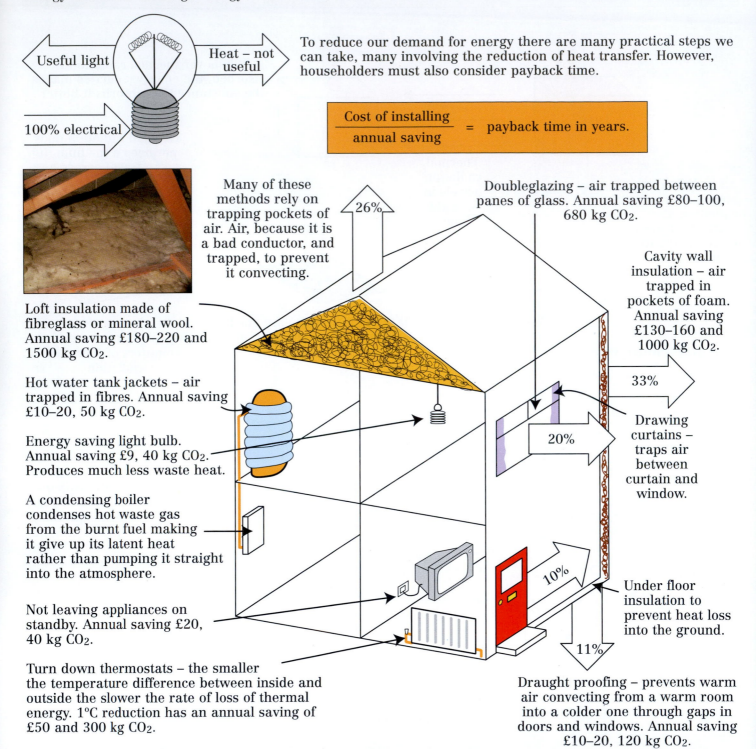

Many of these methods rely on trapping pockets of air. Air, because it is a bad conductor, and trapped, to prevent it convecting.

Loft insulation made of fibreglass or mineral wool. Annual saving £180–220 and 1500 kg CO_2.

Hot water tank jackets – air trapped in fibres. Annual saving £10–20, 50 kg CO_2.

Energy saving light bulb. Annual saving £9, 40 kg CO_2. Produces much less waste heat.

A condensing boiler condenses hot waste gas from the burnt fuel making it give up its latent heat rather than pumping it straight into the atmosphere.

Not leaving appliances on standby. Annual saving £20, 40 kg CO_2.

Turn down thermostats – the smaller the temperature difference between inside and outside the slower the rate of loss of thermal energy. 1°C reduction has an annual saving of £50 and 300 kg CO_2.

Doubleglazing – air trapped between panes of glass. Annual saving £80–100, 680 kg CO_2.

Cavity wall insulation – air trapped in pockets of foam. Annual saving £130–160 and 1000 kg CO_2.

33%

Drawing curtains – traps air between curtain and window.

26%

20%

10%

11%

Under floor insulation to prevent heat loss into the ground.

Draught proofing – prevents warm air convecting from a warm room into a colder one through gaps in doors and windows. Annual saving £10–20, 120 kg CO_2.

Saving energy also reduces carbon dioxide emissions because carbon dioxide is a waste product of burning any fossil fuel, either directly such as gas in a boiler, or indirectly to generate electricity in a power station.

(Data correct (2006) Energy Saving Trust www.est.org.uk.)

Questions
1. A householder could spend £230 on loft insulation that would save £180 in fuel bills each year, or they could spend £75 on draughtproofing and save £20 each year. Which would you recommend they do and why?
2. Why do we talk about wasting energy when physics tells us energy is conserved?
3. Which of the energy saving measures above are free?
4. The annual savings quoted above are both in terms of money and CO_2 saved. Which do you consider to be more important and why?

THERMAL ENERGY Kinetic Model of Gases

The kinetic model of gases is the name we give to the idea that a gas is made up of microscopic particles moving randomly, colliding with each other and the walls of the container.

The pressure is a measure of the rate and force of the collision of the gas molecules with the sides of the container.

The temperature (in Kelvin) is *directly proportional* to the average kinetic energy of the molecules.

As the temperature rises, so does the average kinetic energy and, therefore, speed of the gas particles.

The volume of the particles is insignificant compared to the volume of the container.

At any time there are on average equal numbers of gas particles moving in any direction. This explains why gases exert equal forces in all directions.

Heating the gas increases the average kinetic energy of the molecules, i.e. raises the temperature.

The gas particles travel faster and hit the sides of the container faster, and more often. Therefore, the pressure rises. We assume the particles do not lose any kinetic energy during collisions.

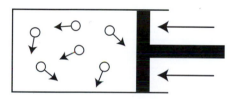

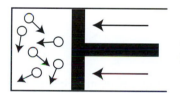

Gas pressure increases.

Gases are very compressible because there are large spaces between the molecules.

Reducing the volume of a container means that the molecules collide with the sides more often as they do not have so far to travel.

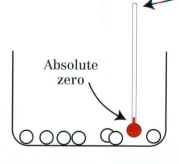

Thermometers read a temperature because gas particles collide with it and transfer kinetic energy to the material of the thermometer.

No pressure exerted at absolute zero, as there are no collisions.

Absolute zero

If the molecules stopped moving, then they would have no kinetic energy – therefore, we would say they had no temperature. This is called *absolute zero* because you cannot get any colder.

THERMAL ENERGY Gas Laws

Measuring the variation of gas pressure with temperature:

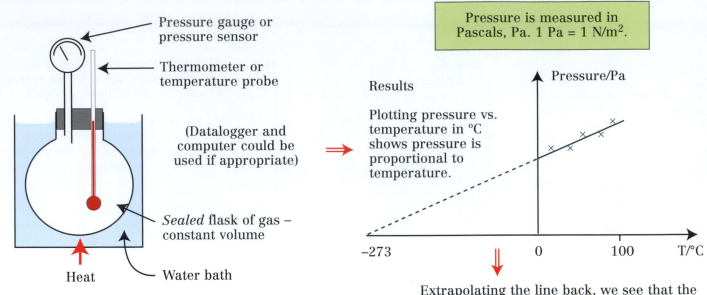

- Pressure gauge or pressure sensor
- Thermometer or temperature probe

(Datalogger and computer could be used if appropriate)

Sealed flask of gas – constant volume

Heat

Water bath

Pressure is measured in Pascals, Pa. 1 Pa = 1 N/m².

Results

Plotting pressure vs. temperature in °C shows pressure is proportional to temperature.

Extrapolating the line back, we see that the pressure would be zero when the temperature is −273°C. This is absolute zero, because all the gas particles would have stopped moving (see p65).

It would make sense to start a temperature scale here – we call this the Kelvin scale.

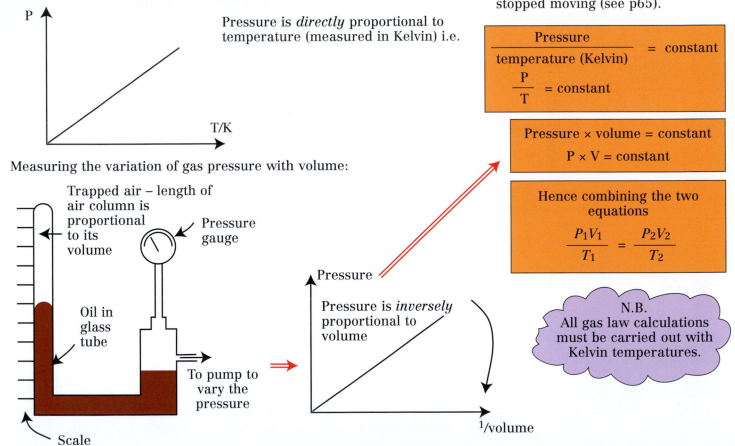

Pressure is *directly* proportional to temperature (measured in Kelvin) i.e.

$$\frac{\text{Pressure}}{\text{temperature (Kelvin)}} = \text{constant}$$

$$\frac{P}{T} = \text{constant}$$

Measuring the variation of gas pressure with volume:

Trapped air – length of air column is proportional to its volume

Pressure gauge

Oil in glass tube

To pump to vary the pressure

Scale

Pressure is *inversely* proportional to volume

$$\text{Pressure} \times \text{volume} = \text{constant}$$

$$P \times V = \text{constant}$$

Hence combining the two equations

$$\frac{P_1 V_1}{T_1} = \frac{P_2 V_2}{T_2}$$

N.B.
All gas law calculations must be carried out with Kelvin temperatures.

Questions
1. The pressure of air in a sealed container at 22°C is 105 000 Pa. The temperature is raised to 85°C. Show that the new pressure is about 130 000 Pa assuming that the volume of the container remains constant.
2. A bubble of air of volume 2 cm³ is released by a deep-sea diver at a depth where the pressure is 420 000 Pa. Assuming the temperature remains constant show that its volume is 8 cm³ just before it reaches the surface where the pressure is 105 000 Pa.
3. A sealed syringe contains 60 cm³ of air at a pressure of 105 000 Pa and at 22°C. The piston is pushed in rapidly until the volume is 25 cm³ and the pressure is 315 000 Pa. Show that the temperature of the gas rises to about 95°C.
4. When a star forms a gas cloud in space is attracted together by gravity compressing it. As the volume of the gas reduces what happens to its pressure and hence temperature?

RADIOACTIVITY Atomic Structure

All atoms have the same basic structure:

Orbiting electrons (negative charge)

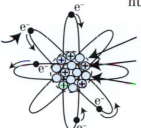

Electrons are held in orbit around the nucleus by electrostatic attraction.

Nucleus, comprising of:
- Protons (positive charge)
- and neutrons (no charge)

} Nucleons as they make up the nucleus.

In all atoms the number of protons = number of electrons. This makes atoms uncharged, or *neutral*.

Naming atoms:

Atomic (proton) number Z = number of protons in the nucleus

Mass (nucleon) number A = total number of protons plus neutrons in the nucleus

$_Z^A X$

Symbol for the element

Each element has a unique number of protons. Therefore, the atomic number uniquely identifies the element.

Some atoms of the same element have different numbers of neutrons.

 $_6^{12}C$ $_6^{13}C$ $_6^{14}C$

E.g. all these atoms are carbon as they all have 6 protons, but they have different numbers of neutrons. They are called isotopes of carbon.

> Isotopes are always the same element, i.e. same atomic number but have different numbers of neutrons (and so mass number).

What is Radioactivity?

Some elements give out random bursts of radiation. Each individual nucleus can only do this once, and when it has happened, it is said to have decayed. As even a tiny sample of material contains billions of atoms, many bursts of radiation can be emitted before all the nuclei have decayed.

Ionizing – it can knock electrons out of other atoms.

The emission of this radiation is random but it decreases over time, sometimes slowly, sometimes very quickly as the number of nuclei left to decay decreases.

The atom left behind is now charged and is called an *ion*. It does not have equal numbers of protons and electrons.

Elements that behave like this are called *radioactive*.

We can measure the radioactivity as the number of decays (and, therefore, bursts of radiation emitted) per second.

> 1 decay per second = 1 Becquerel, Bq

The relative masses of protons, neutrons, and electrons and their relative electric charges are:

	Mass	Charge
Proton	1	+1
Neutron	1	0
Electron	$\dfrac{1}{1870}$	−1

Questions

1. Copy and complete the table.

	No. of protons	No. of electrons	No. of neutrons
Carbon $_6^{12}C$			
Barium $_{56}^{137}Ba$			
Lead $_{=}Pb$		82	125
Iron ^{56}Fe	26		
Hydrogen $_1^1H$			
Helium $_2^4He$			
Helium $_2^3He$			
Element X $_Z^A X$			

2. a Draw a diagram to show all the protons and neutrons in the nuclei of $_{17}^{35}Cl$ and $_{17}^{37}Cl$.
   b. What word do we use to describe these two nuclei?
   c. Why is there no difference in the way the two types of chlorine atoms behave in chemical reactions?
   d. If naturally occurring chlorine is 75% $_{17}^{35}Cl$ and 25% $_{17}^{37}Cl$ explain why on a periodic table it is recorded as $_{17}^{35.5}Cl$?

3. What is a Becquerel?

4. If ionizing radiation knocks electrons out of atoms, will the ions left behind be positively or negatively charged? Why?

5. Explain what you understand by the term 'radioactive element'.

RADIOACTIVITY A History of Our Understanding of the Atom

In 1803, John Dalton noted that chemical compounds always formed from the same ratio of elements, suggesting particles were involved. He called these atoms from the Greek, meaning indivisible.

J.J. Thomson (1897) discovered the electron, a particle that could be knocked out of an atom. He suggested a 'plum pudding' model of the atom.

Rutherford, Geiger, and Marsden investigated this in 1910. They decided to probe the nucleus further with alpha particles. These are particles with two positive charges, which they considered to be like little bullets.

Incoming electron

Extra electron knocked out

e⁻

e⁻

The model was so named as the electrons appeared like the fruit in a pudding.

Electrons

Sphere of positive charge

1. Detector detects the alpha particles that have travelled through the foil. It can be moved to any angle round the foil so that the number of alpha particles in any direction can be recorded.

2. The majority of alpha particles travelled through the foil with very little change in direction.

3. A *very* small number were turned through angles greater than 90°.

5. Rutherford proposed the nuclear model.

Source of alpha particles

About 10^{-14} m in diameter

Inner chamber evacuated.

Thin gold foil

4. Plum pudding model cannot explain this since as the positive and negative charges were reasonably evenly distributed no alpha particles should get scattered through large angles.

Protons in nucleus with orbiting electrons

Alpha particle (2 protons, charge +2)

Large scattering angle when an alpha particle passes close to nucleus, small when far away.

Gold foil

Gold nucleus (79 protons, charge +79)

Rarity of large angle of scatter tells us the nucleus is very small.

Repulsive electrostatic force changes direction of the alpha particle.

Summary
The alpha scattering experiment proves that:
1. Atoms have massive, positively charged nuclei.
2. The majority of the mass of the atom is the nucleus.
3. Electrons orbit outside of the nucleus. Most of the atom is empty space.

Bohr further developed the atomic model by suggesting that the electrons were arranged in energy levels around the nucleus.

Kinetic energy is transferred to potential energy in the electric field round the nucleus as the alpha particle does work against the repulsive force. This is returned to kinetic energy on leaving the region near the nucleus.

- The larger the charge on the nucleus the greater was the angle of scatter.
- The thicker the foil the greater the probability that an alpha particle passes close to a nucleus.
- Slower alpha particles remain in the field around the nucleus for longer – increases the angle of scattering.

To move up a level it has to absorb precisely the right amount of energy from an electromagnetic wave.

If an electron moved down a level, it has to get rid of some energy in the form of an electromagnetic wave.

Questions
1. List the main conclusions of the alpha scattering experiment.
2. What evidence did Thomson have for the plum pudding model?
3. Suggest why the alpha scattering apparatus has to be evacuated (have all the air taken out of it).
4. Suggest why the gold foil used in the alpha scattering experiment needs to be very thin.
5. The diameter of an atom is about 10^{-10} m and of a gold nucleus 10^{-14} m. Show that the probability of directly hitting a nucleus with an alpha particle is about 1 in 108. What assumptions have you made?

RADIOACTIVITY Background Radiation

Radioactive elements are naturally found in the environment and are continually emitting radiation. This naturally occurring radiation is called *background radiation*, which we are all exposed to throughout our lives.

Background radiation comes from a number of sources. (Note that these are averaged across the population and may differ for different groups, for example depending on any medical treatment you may have, or whether you make many aeroplane flights.)

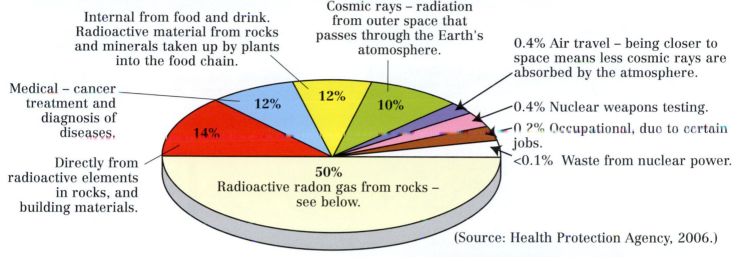

Internal from food and drink. Radioactive material from rocks and minerals taken up by plants into the food chain.

Cosmic rays – radiation from outer space that passes through the Earth's atomosphere.

0.4% Air travel – being closer to space means less cosmic rays are absorbed by the atmosphere.

Medical – cancer treatment and diagnosis of diseases.

0.4% Nuclear weapons testing.

0.2% Occupational, due to certain jobs.

<0.1% Waste from nuclear power.

Directly from radioactive elements in rocks, and building materials.

14%

12%

12%

10%

50%
Radioactive radon gas from rocks – see below.

(Source: Health Protection Agency, 2006.)

One of the major sources of background radiation is radon gas. This is produced by minute amounts of uranium, which occurs naturally in rocks, and is present in all parts of the country. It disperses outdoors so is only a problem if trapped inside a building. Exposure to high levels of radon can lead to an increased risk of lung cancer.

Since we all inhale radon throughout our lives it accounts for about half our annual radiation dose in the UK.

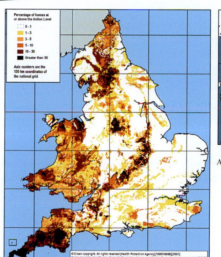

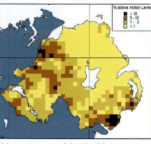

Maps courtesy of the Health Protection Agency and the British Geological Survey

For more information go to:
www.ukradon.co.uk

Geological conditions in some areas produce higher than average radon concentrations as shown in the map.

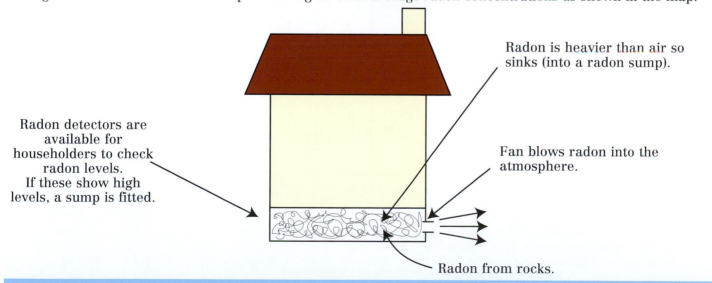

Radon is heavier than air so sinks (into a radon sump).

Radon detectors are available for householders to check radon levels. If these show high levels, a sump is fitted.

Fan blows radon into the atmosphere.

Radon from rocks.

Questions
1. Make a list of sources of background radiation.
2. Give at least two reasons why the percentages shown above in the sources of background radiation are only averages and will differ for different people.
3. On average what percentage of the total background radiation is man-made?
4. Should we worry about background radiation?

RADIOACTIVITY Three Types of Nuclear Radiation

There are three types of radiation emitted by radioactive materials. They are all emitted from unstable *nuclei*:

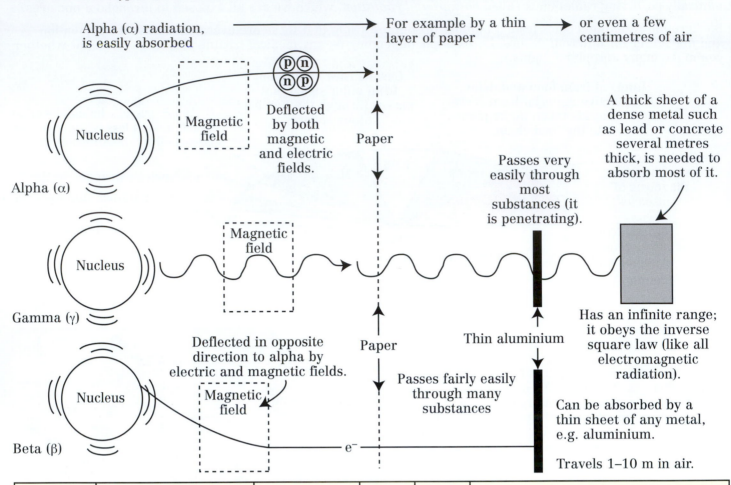

Name	Identity		Mass	Charge	
Alpha (α)	Helium *nucleus*	$^4_2$He	4	+2	Massive and highly charged. Therefore, interacts strongly with other matter causing ionization, and loses energy rapidly. Easily stopped and short range
Beta (β)	Fast moving electron ejected from the *nucleus*. Note that it is not an atomic orbital electron	e⁻	$\dfrac{1}{1870}$	−1	Nearly 8000 × less massive than alpha and only half the charge. Therefore, does not interact as strongly with other matter causing less ionization, and loses energy more gradually. Harder to stop and has a longer range
Gamma (γ)	Electromagnetic wave		0	0	No mass or charge so only weakly interacts with matter. Therefore, very difficult to stop

Questions
1. Describe the differences between alpha, beta, and gamma radiation. What materials will stop each one?
2. Alpha and beta particles are deflected in both electric and magnetic fields but gamma is not. Explain why. Why are alpha and beta deflected in opposite directions?
3. A student has a radioactive source. When the source is placed 1 cm in front of a GM tube connected to a ratemeter it counts 600 counts per minute.
   - Moving the source back to 10 cm the count drops to 300 counts per minute.
   - Replacing the source at 1 cm and inserting 2 mm thickness of aluminium foil gives 300 counts per minute.
   - Moving the source back to 5 cm and inserting 2 cm of lead gives 150 counts per minute.
   Explain how you know what type(s) of radiation the source emits.
4. Many smoke alarms contain a small radioactive source emitting alpha particles. This is inside an aluminium box, and placed high on a ceiling. Use the properties of alpha particles to explain why smoke alarms do not pose any health risk.

RADIOACTIVITY Radioactive Decay and Equations

Most nuclei never change; they are stable. Radioactive materials contain unstable nuclei. These can break up and emit radiation. When this happens, we say the nucleus has *decayed*. The result for alpha and beta decay is the nucleus of a different element. For gamma decay, it is the same element but it has less energy.

Alpha decay

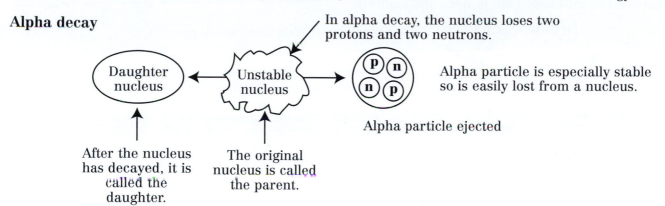

In alpha decay, the nucleus loses two protons and two neutrons.

Alpha particle is especially stable so is easily lost from a nucleus.

Alpha particle ejected

Daughter nucleus

Unstable nucleus

After the nucleus has decayed, it is called the daughter.

The original nucleus is called the parent.

Mass number decreases by 4 (2 protons + 2 neutrons lost). Atomic number decreases by 2 (2 protons lost).

Atomic number

$$_{Z}^{A}X \rightarrow\ _{(Z-2)}^{(A-4)}Y + _{2}^{4}He \qquad \text{Or} \qquad _{Z}^{A}X \rightarrow\ _{(Z-2)}^{(A-4)}Y + _{2}^{4}\alpha$$

Beta decay

Beta-minus $\ p \leftarrow n \rightarrow e^-$

Neutron becomes a proton and electron.

Daughter nucleus has one more proton than the parent so the atomic number increases by one.

Overall number of protons plus neutrons is unchanged so the mass number does not change.

$$_{Z}^{A}X \rightarrow\ _{(Z+1)}^{A}Y + _{-1}^{0}e^-$$

Or

$$_{Z}^{A}X \rightarrow\ _{(Z+1)}^{A}Y + _{-1}^{0}\beta^-$$

Beta-plus $\ n \leftarrow p \rightarrow e^+$

Proton becomes a neutron and a positron (an anti-electron with all the same properties as an electron but the opposite charge).

Daughter nucleus has one less proton than the parent so the atomic number decreases by one.

Overall number of protons plus neutrons is unchanged so the mass number does not change.

$$_{Z}^{A}X \rightarrow\ _{(Z-1)}^{A}Y + _{+1}^{0}e^+$$

Or

$$_{Z}^{A}X \rightarrow\ _{(Z-1)}^{A}Y + _{+1}^{0}\beta^+$$

Gamma decay

Often after either alpha or beta decay the nucleons have an excess of energy. By rearranging the layout of their protons and neutrons, they reach a lower energy state and the excess energy is emitted in the form of a gamma ray.

$$_{Z}^{A}X \rightarrow\ _{Z}^{A}X + \gamma$$

Rules for nuclear equations

The total mass number must be the same on both sides of the equation.

The total atomic number on both sides of the equation must be the same.

The total charge must be the same on both sides of the equation.

Questions

Copy and complete the following nuclear equations:

1. $_{84}^{215}Po \rightarrow\ _{82}^{211}Pb +$ ___.
2. $_{90}^{228}Th \rightarrow\ _{—}^{—}Ra + _{2}^{4}\alpha$.
3. $_{82}^{214}Pb \rightarrow\ _{83}^{214}Bi +$ ___.
4. $_{8}^{15}O \rightarrow\ _{7}^{15}N +$ ___.

5. $_{—}^{—}Si \rightarrow\ _{13}^{27}Al + _{+1}^{0}$___⁺.
6. $_{—}^{238}U \rightarrow\ _{90}^{—}Th + _{2}^{4}\alpha$.
7. $_{33}^{74}As\ \cdot\ _{—}^{—}Se + _{—}^{0}$___.
8. $_{89}^{227}Ac \rightarrow\ _{87}^{—}Fr +$ ___.

RADIOACTIVITY N/Z Curve

Nuclei have positive charge due to the protons in them. All the protons repel, so why does the nucleus not explode?

There is another force acting called the *strong nuclear force*.
This acts between all nucleons, both protons and neutrons.

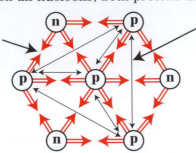

Strong nuclear attraction only acts between adjacent nucleons.

Electrostatic repulsion between all protons.

For small nuclei, a proton:neutron ratio of 1:1 is sufficient for the strong nuclear force to balance the electrostatic force. For larger nuclei, we need more neutrons to provide extra strong nuclear force, without increasing the electrostatic repulsion, so the ratio rises to 1.6:1.

Plotting the number of protons vs. number of neutrons in stable nuclei gives this graph.

For elements where $Z > 80$ these decay by α decay. →

Alpha particles consist of two protons and two neutrons.
Therefore, the atomic number falls by two and the mass number by four.

$$_Z^A X \rightarrow {}_{(Z-2)}^{(A-4)} Y + {}_2^4 He$$

N.B. Remember alpha particle is $_2^4 He$.

These isotopes need to gain protons and lose neutrons to move towards the line of stability. They have too much strong nuclear force and not enough electrostatic force. β^- *decay* allows this to happen.
A neutron turns into a proton and an electron. The equations for this process are:

$n \rightarrow p + e^-$

Overall $_Z^A X \rightarrow {}_{(Z+1)}^{A} Y + {}_{-1}^{0} \beta^-$.

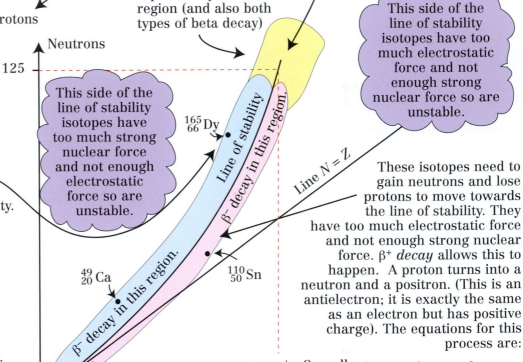

Alpha decay in this region (and also both types of beta decay)

If a nucleus has a proton:neutron mixture close to this line is stable and does not decay.

This side of the line of stability isotopes have too much electrostatic force and not enough strong nuclear force so are unstable.

This side of the line of stability isotopes have too much strong nuclear force and not enough electrostatic force so are unstable.

Neutrons

125

$_{66}^{165}$Dy

Line of stability

β^- decay in this region.

Line $N = Z$

$_{20}^{49}$Ca

β^- decay in this region.

$_{50}^{110}$Sn

These isotopes need to gain neutrons and lose protons to move towards the line of stability. They have too much electrostatic force and not enough strong nuclear force. β^+ *decay* allows this to happen. A proton turns into a neutron and a positron. (This is an antielectron; it is exactly the same as an electron but has positive charge). The equations for this process are:

$p \rightarrow n + e^+$ Overall $_Z^A X \rightarrow {}_{(Z-1)}^{A} Y + {}_{+1}^{0} \beta^+$.

6

$_{10}^{18}$Ne

6

80

Protons

N.B. remember the beta particle is an electron.

E.g.

$_{20}^{49}$Ca $\rightarrow$ $_{21}^{49}$Sc $+ {}_{-1}^{0}\beta^-$.

$_{66}^{165}$Dy $\rightarrow {}_{67}^{165}$Ho $+ {}_{-1}^{0}\beta^-$.

↑   ↖
Parent   Daughter

E.g.

$_{50}^{110}$Sn $\rightarrow {}_{49}^{110}$In $+ {}_{+1}^{0}\beta^+$.

$_{10}^{18}$Ne $\rightarrow {}_{9}^{18}$F $+ {}_{+1}^{0}\beta^+$.

Questions
1. Explain why proportionately more neutrons are needed in larger nuclei?
2. Using the graph above, calculate the ratio $Z:N$ when $Z = 6$ and when $Z = 80$. Comment on your answer. Why does the line on the graph curve away from the line $Z = N$?
3. What type of decay occurs in isotopes with too much strong nuclear force? How do these changes help the nucleus to become more stable?
4. Repeat question 3 for isotopes with too much electrostatic force.
5. Nuclei do not contain electrons, so where does the electron emitted from a nucleus in beta-minus decay come from?
6. Balance the equation $_6^{11} C \rightarrow {}^{11} B + \_\_\_\_$. (Hint: are there too many protons or too many neutrons in the carbon nucleus?), hence will β^+ or β^- decay occur?

RADIOACTIVITY Fundamental Particles

A fundamental particle is one that cannot be split into anything simpler.

The word atom means 'indivisible' because scientists once thought atoms were fundamental particles.

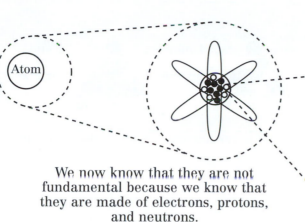

We now know that they are not fundamental because we know that they are made of electrons, protons, and neutrons.

Similar experiments to Rutherford's alpha scattering using electrons fired at protons and neutrons reveals that they are made up of smaller particles – *quarks*.

Scientists now think that quarks, together with electrons and **positrons** are examples of fundamental particles.

There are actually six types of quark given odd names. They also have fractional charges as shown below.

An example of antimatter. All particles have antiparticles; they are identical in mass but opposite in charge. Our Universe is made of matter. Antimatter is made in particle accelerators or as the result of some nuclear processes such as beta-plus decay.

Up	Charge	Charm	Charge	Top	Charge
u	$+^2/_3$	c	$+^2/_3$	t	$+^2/_3$
Down	Charge	Strange	Charge	Bottom	Charge
d	$-^1/_3$	s	$-^1/_3$	b	$-^1/_3$

Normally we are not allowed fractional charges, but quarks never occur on their own, only in combinations that add up to a whole charge.

Beta decay

In beta decay, one of the up quarks changes to a down quark or *vice versa*.

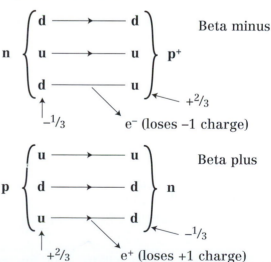

Protons and neutrons are made of just two types of quark, the up and the down. Other particles have to be created in special machines called particle accelerators.

Proton – two up and one down quarks.

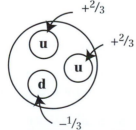

Charge = $(+^2/_3) + (+^2/_3) + (-^1/_3) = +1$

Neutron – one up and two down quarks.

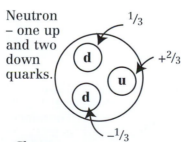

Charge = $(+^2/_3) + (-^1/_3) + (-^1/_3) = 0$

RADIOACTIVITY Half-Life

Most types of nuclei never change; they are stable. However, radioactive materials contain unstable nuclei. The nucleus of an unstable atom can break up (decay) and when this happens, it emits radiation.

A nucleus of a different element is left behind.

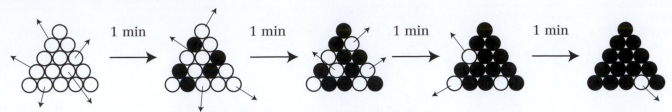

As time goes by radioactive materials contain fewer and fewer unstable atoms and so become less and less radioactive and emit less and less radiation.

There is no way of predicting when an individual nucleus will decay; it is a completely random process. A nucleus may decay in the next second or not for a million years. This means it is impossible to tell how long it will take for all the nuclei to decay.

Like throwing a die, you cannot predict when a six will be thrown. However, given a very large number of dice you can estimate that a certain proportion, $\frac{1}{6}$th, will land as a six.

We define *activity* as the number of nuclei that decay per second (N.B. 1 decay per second = 1 Bq). The time it takes for the activity of a radioactive material to halve (because half of the unstable nuclei that were originally there have decayed) is called the **half-life**.

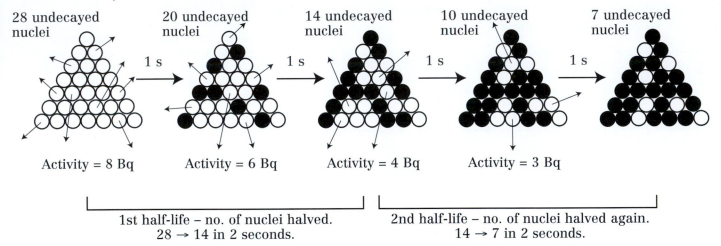

28 undecayed nuclei	20 undecayed nuclei	14 undecayed nuclei	10 undecayed nuclei	7 undecayed nuclei
Activity = 8 Bq	Activity = 6 Bq	Activity = 4 Bq	Activity = 3 Bq	

1st half-life – no. of nuclei halved.
28 → 14 in 2 seconds.

2nd half-life – no. of nuclei halved again.
14 → 7 in 2 seconds.

We see the activity falling as there are fewer nuclei available to decay. However, note that the time taken to halve is independent of the number of nuclei, in this case 2 seconds. Half-lives are unique to each individual isotope and range from billions of years to fractions of a second.

The half-life of a radioactive isotope is formally defined as:

> 'The time it takes for half the nuclei of the isotope in a sample to decay, or the time it takes for the count rate from a sample containing the isotope to fall to half its initial level.'

Calculations

1. Numerically e.g. a radioisotope has an activity of 6400 Bq and a half-life of 15 mins.

After 15 mins the activity will be $\frac{6400 \text{ Bq}}{2} = 3200 \text{ Bq}$.

After 30 mins the activity will be $\frac{3200 \text{ Bq}}{2} = 1600 \text{ Bq}$.

After 45 mins the activity will be $\frac{1600 \text{ Bq}}{2} = 800 \text{ Bq}$.

After 1 hour the activity will be $\frac{800 \text{ Bq}}{2} = 400 \text{ Bq}$.

Alternatively, consider the number of half-lives, e.g. $1\frac{1}{2}$ hrs = 6 × 15 mins = 6 half-lives. Therefore

$$\text{activity} = \frac{\text{original activity}}{(2 \times 2 \times 2 \times 2 \times 2 \times 2)}$$

(i.e. divide by 2, six times)

$$= \frac{\text{original activity}}{2^6}$$

In general,

$$\text{activity} = \frac{\text{original activity}}{2^{\text{no. of half-lives}}}$$

Therefore after 6 half-lives, in this case,

$$\text{activity} = \frac{6400 \text{ Bq}}{2^6} = 100 \text{ Bq}.$$

2. Graphically A graph of activity vs. time can be plotted from experimental measurements. We must remember to subtract the background count from the actual count to find the count due to the source alone. We call this the *corrected count rate*.

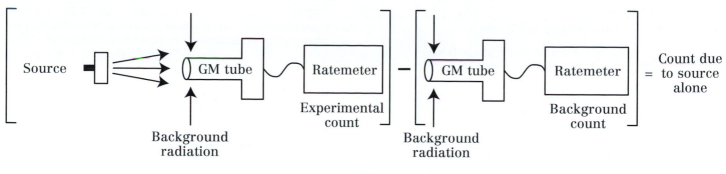

Count due
= to source
alone

Experimental count — Background count

Background radiation   Background radiation

Results.

As long as the sample is large enough, this curve is smooth, because although the process is random, probability tells us half the atoms will decay in a certain time, but not which.

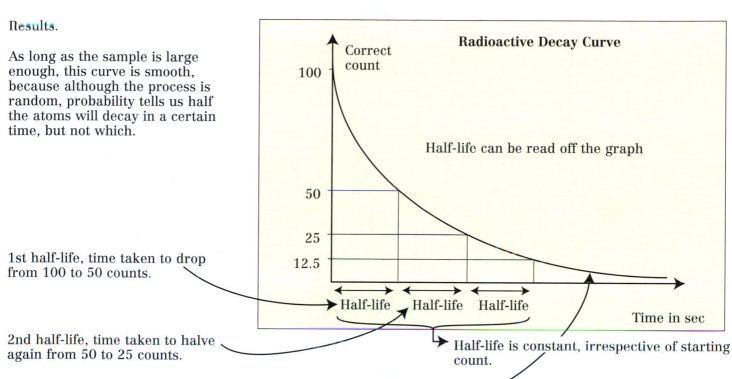

Radioactive Decay Curve

Correct count

Half-life can be read off the graph

1st half-life, time taken to drop from 100 to 50 counts.

Half-life   Half-life   Half-life

Time in sec

2nd half-life, time taken to halve again from 50 to 25 counts.

Half-life is constant, irrespective of starting count.

Nuclear radiation never completely dies away, but eventually drops to a negligible level, close to the background. At this point, a source is considered safe. Consideration of half-life therefore, has importance when considering which isotopes to use for various applications and the disposal of radioactive waste – see section on applications of radioactivity.

Questions

1. What is the activity of a radioactive source?
2. Write down a definition of half-life. Suggest why we can measure the half-life of a substance, but not its 'full life' (i.e. the time for all the atoms to decay).
3. $^{99}_{43}$Tc (Technetium) has a half-life of 6 hrs. A sample of technetium has an initial count rate of 128 000 Bq
   i. What will the count rate be after: a. 6 hrs? b. 18 hrs?
   ii. How many hours will it take the count rate to fall to: a. 32 000 Bq? b. 8000 Bq? c. 1000 Bq?
4. A student has a sample of $^{137}_{56}$Ba (Barium). They record the count rate every 60 s and record the following results:

Time in seconds	0	60	120	180	240	300	360	420	480	540	600	660	720
Count rate (decays/s)	30.8	23.8	18.4	14.2	11.1	8.7	6.9	5.4	4.4	3.5	2.9	2.4	2.0

The background count rate, with no source present, was 0.8 counts per second.
   a. Copy the table and include a row for the corrected count rate.
   b. Draw a graph of count rate vs. time and use it to show that the half-life is approximately 156 s.
   c. Do you think this isotope would present significant disposal problems, why or why not?
5. A student has a sample of radioactive material. In one lesson the activity recorded was 2000 Bq. The next day, at the same time, the count rate was just over 500 Bq. Which of the following isotopes is the sample most likely to be?
   a. $^{135}_{53}$I (iodine) half-life – 6.7 hrs.   c. $^{42}_{19}$K (potassium) half-life = 12.5 hrs.
   b. $^{87}_{38}$Sr (strontium) half-life = 2.9 hrs.   d. $^{187}_{74}$W (tungsten) half-life = 24 hrs.

RADIOACTIVITY Is Radiation Dangerous?

All nuclear radiation is ionizing. It can knock electrons out of atoms, or break molecules into bits. If these molecules are part of a living cell, this may kill the cell.

If the molecule is DNA, the damage caused by the radiation may affect the way it replicates. This is called *mutation*. Sometimes this leads to *cancer*.

Alpha particles are heavy and highly charged, and interact strongly with atoms. They can travel only very short distances and are easily stopped. They cannot penetrate human skin. Alpha emitters are only dangerous when inhaled, ingested, injected, or absorbed through a wound.

Radiation dose is measured in Sieverts. This unit measures the amount of energy deposited in the tissue by the radiation, and takes account of the type of radiation, because some particles are more effective at damaging cells than others. It is a measure of the possible harm done to your body.

Beta particles are also charged, but interact less strongly than alpha particles, so travel further and penetrate more: they can penetrate skin. Clothing provides some protection. They can cause radiation burns on prolonged exposure but are hazardous to internal organs only when inhaled, ingested, injected, or absorbed.

Gamma rays are uncharged, so do not interact directly with atoms, and travel many metres in air. They easily penetrate the human body, causing organ damage. Their effects can be reduced by concrete or lead shielding.

Many people work with radiation, e.g. radiologists in hospitals, and nuclear power plant workers. Their exposure is carefully recorded. They wear a film badge, which becomes gradually more fogged, depending on how much exposure they have had. If their exposure is too high in a set period, they will usually be given other jobs away from radiation sources, temporarily.

Irradiation occurs when the emitted radiation hits an object. Moving away will reduce the exposure.

Radioactive materials have to be handled safely. Various precautions to adopt include:

- Keeping source as far from body as possible – usually using tongs.

- Protective clothing – usually only for highly active sources.

- Keeping exposure time as short as possible.

- Keeping the source in appropriate storage, usually shielded, e.g. lead, and labelled.

Something is *contaminated* if the radioactive atoms are in contact with it. Moving away will spread the contamination.

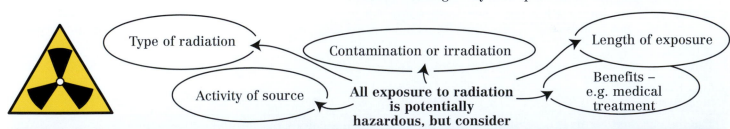

Type of radiation

Contamination or irradiation

Length of exposure

Activity of source

All exposure to radiation is potentially hazardous, but consider

Benefits – e.g. medical treatment

Questions
1. Explain which type of radiation is most harmful:
   a. Outside the body.
   b. Inside the body.
2. Explain the difference between contamination and irradiation. Which would you consider a more serious problem?
3. How does nuclear radiation cause damage to living tissues?
4. What is a Sievert?
5. Explain three precautions you should take if you had to handle a low activity radioactive source.

RADIOACTIVITY Nuclear Fission

Nuclear fission is the splitting of an atomic nucleus.

A large parent nucleus, such as 235-uranium or 239-plutonium, splits into two smaller daughter nuclei, of approximately equal size. This process also releases energy (heat) which can be used to generate electricity (see p111). Normally, this will happen spontaneously but can be speeded up by inducing fission.

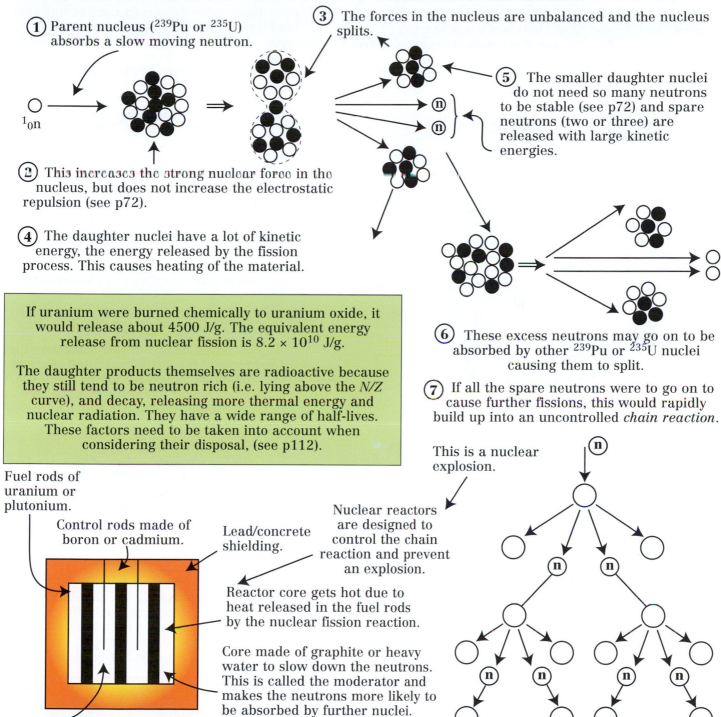

① Parent nucleus (^{239}Pu or ^{235}U) absorbs a slow moving neutron.

$^1_0$n

② This increases the strong nuclear force in the nucleus, but does not increase the electrostatic repulsion (see p72).

④ The daughter nuclei have a lot of kinetic energy, the energy released by the fission process. This causes heating of the material.

③ The forces in the nucleus are unbalanced and the nucleus splits.

⑤ The smaller daughter nuclei do not need so many neutrons to be stable (see p72) and spare neutrons (two or three) are released with large kinetic energies.

⑥ These excess neutrons may go on to be absorbed by other ^{239}Pu or ^{235}U nuclei causing them to split.

⑦ If all the spare neutrons were to go on to cause further fissions, this would rapidly build up into an uncontrolled *chain reaction*.

This is a nuclear explosion.

Nuclear reactors are designed to control the chain reaction and prevent an explosion.

If uranium were burned chemically to uranium oxide, it would release about 4500 J/g. The equivalent energy release from nuclear fission is 8.2×10^{10} J/g.

The daughter products themselves are radioactive because they still tend to be neutron rich (i.e. lying above the *N/Z* curve), and decay, releasing more thermal energy and nuclear radiation. They have a wide range of half-lives. These factors need to be taken into account when considering their disposal, (see p112).

Fuel rods of uranium or plutonium.

Control rods made of boron or cadmium.

Lead/concrete shielding.

Reactor core gets hot due to heat released in the fuel rods by the nuclear fission reaction.

Core made of graphite or heavy water to slow down the neutrons. This is called the moderator and makes the neutrons more likely to be absorbed by further nuclei.

Control rods absorb neutrons before they can cause further fissions.

and so on . . .

Lowering the control rods absorbs more neutrons and slows the reaction, raising the control rods speeds it up.

Questions

1. Balance this equation, a fission reaction of uranium producing the daughter nuclei barium and krypton.
   $^{235}_{92}$U + $^1_0$n → ____$_{56}$Ba + ^{90}___ Kr + 2 $^1_0$n.
2. In what form is the majority of the energy released by a nuclear reaction?
3. Why do the products of fission reactions need careful handling?
4. How do the control rods in a reactor control the rate of the nuclear reaction?
5. For a stable chain reaction, neither speeding up nor slowing down, suggest how many neutrons from each fission should go on to cause a further fission.
6. Use the data above to show that the energy released from the fission of 1 g of ^{235}U is about 20 million times as much as when the same gram is burnt in oxygen to form uranium oxide.

RADIOACTIVITY   Nuclear Fusion

Nuclear fusion is the joining of two light nuclei to form a heavier nucleus. It is the process by which energy is released in stars.

In the nucleus, the STRONG NUCLEAR FORCE attracts protons and neutrons together; it is stronger than the ELECTROSTATIC REPULSION between the protons but it is a very short-range force.

To fuse two nuclei they must be brought very close together so the strong nuclear force can bind their protons and neutrons together.

To do this you have to overcome the electrostatic repulsion between the nuclei.

This means that the gas containing the nuclei has to be very hot, dense, and under high pressure.

The gas is so hot that none of the electrons now orbits the nuclei. This is called plasma.

Therefore, the nuclei have to travel very fast so they have a lot of kinetic energy to do work against the repulsive force.

When the nuclei join, energy is released as the kinetic energy of the product nucleus.

The nucleus formed has less mass than the total mass of the nuclei that fused to create it. The missing mass (or mass defect) has been converted to energy by Einstein's famous relationship

$$\Delta E = \Delta mc^2$$

ΔE = energy released in J
Δm = mass loss in kg
c = speed of light = 3×10^8 m/s

This is very difficult to do on Earth as this plasma would melt any container. Confining plasma is a major area of research because for the same mass of fuel, fusion of hydrogen to helium releases much more energy than fission and is the reaction occurring in the core of stars. We have a plentiful supply of hydrogen in water on Earth and the products are not polluting.

Scientists still have not achieved the process under control. They can do it where the reaction is explosive, in a hydrogen bomb. Some scientists once claimed they could do fusion at room temperature, but no one has been able to repeat this.

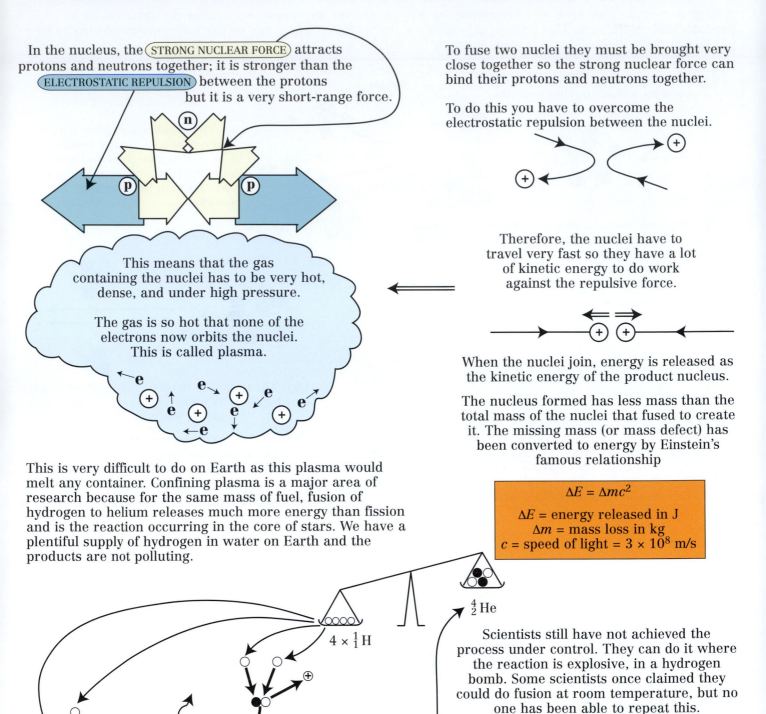

$4 \times {}^1_1$H

${}^4_2$He

Gamma rays

H⁺

2 protons recycled

${}^4_2$He nucleus

ENERGY RELEASED

Positron (proton converted to neutron by β-plus decay)

H⁺

Key

○ H⁺ proton

○● Deuterium nucleus (1n + 1p)

●○ ${}^3_2$He nucleus

⊕ Positron (β⁺ particle)

Questions
1. Explain the differences between nuclear fission and fusion.
2. What are the two forces that must be kept in balance in a stable nucleus?
3. What is plasma?
4. Why does fusion require such high temperatures and what problems may occur as a result?
5. Explain why scientists are working hard to achieve controlled fusion on Earth.
6. A helium-4 nucleus is only 99.3% of the mass of the 4 hydrogen nuclei from which it was formed. The other 0.7% of its mass is converted into energy. Use Einstein's equation $\Delta E=\Delta mc^2$ to show that the energy released from the fusion of 1 kg of hydrogen nuclei, is about 6.3×10^{14} J (c = speed of light = 3×10^8 m/s).

APPLICATIONS OF PHYSICS

The previous pages have outlined some of the main ideas that physicists believe. Physicists hold these ideas because they have collected evidence.

How Science Works

The remainder of this book outlines some of the ways that these ideas have been put to use. The link between these two aspects is how science works.

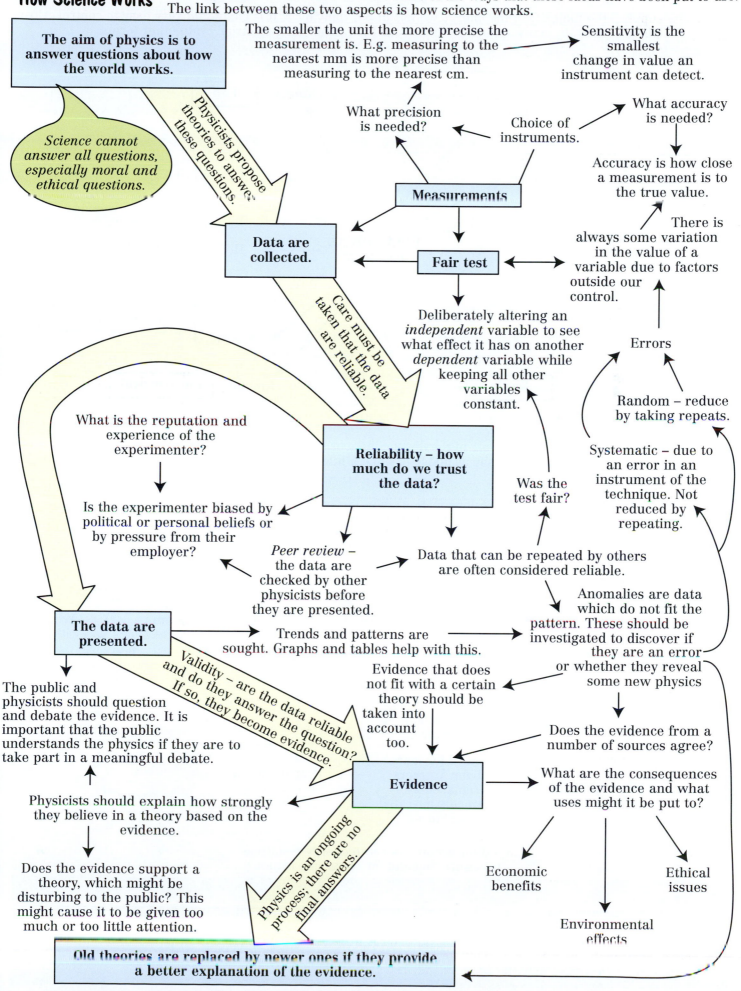

The aim of physics is to answer questions about how the world works.

Science cannot answer all questions, especially moral and ethical questions.

Physicists propose theories to answer these questions.

The smaller the unit the more precise the measurement is. E.g. measuring to the nearest mm is more precise than measuring to the nearest cm.

Sensitivity is the smallest change in value an instrument can detect.

What precision is needed?

Choice of instruments.

What accuracy is needed?

Accuracy is how close a measurement is to the true value.

Data are collected.

Measurements

Fair test

There is always some variation in the value of a variable due to factors outside our control.

Deliberately altering an *independent* variable to see what effect it has on another *dependent* variable while keeping all other variables constant.

Errors

Random – reduce by taking repeats.

Care must be taken that the data are reliable.

What is the reputation and experience of the experimenter?

Is the experimenter biased by political or personal beliefs or by pressure from their employer?

Reliability – how much do we trust the data?

Was the test fair?

Systematic – due to an error in an instrument of the technique. Not reduced by repeating.

Peer review – the data are checked by other physicists before they are presented.

Data that can be repeated by others are often considered reliable.

Anomalies are data which do not fit the pattern. These should be investigated to discover if they are an error or whether they reveal some new physics

The data are presented.

Trends and patterns are sought. Graphs and tables help with this.

Validity – are the data reliable and do they answer the question? If so, they become evidence.

Evidence that does not fit with a certain theory should be taken into account too.

Does the evidence from a number of sources agree?

The public and physicists should question and debate the evidence. It is important that the public understands the physics if they are to take part in a meaningful debate.

Evidence

What are the consequences of the evidence and what uses might it be put to?

Physicists should explain how strongly they believe in a theory based on the evidence.

Physics is an ongoing process; there are no final answers.

Economic benefits

Ethical issues

Does the evidence support a theory, which might be disturbing to the public? This might cause it to be given too much or too little attention.

Environmental effects

Old theories are replaced by newer ones if they provide a better explanation of the evidence.

THE SUPPLY AND USE OF ELECTRICAL ENERGY

Examples of Energy Transformations Involving Electrical Devices and the Impact of Electricity on Society

Electricity supplies the majority of the energy we use in our daily lives. It is clean and very easy to control. Most houses contain many appliances that work by transforming electricity into other forms.

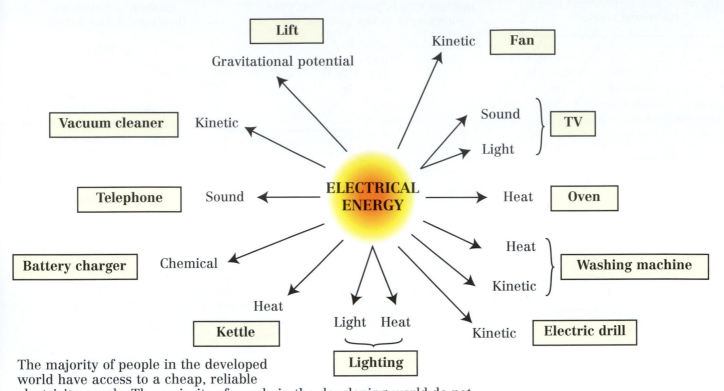

The majority of people in the developed world have access to a cheap, reliable electricity supply. The majority of people in the developing world do not.
The fact that we have electricity so easily available has had a huge impact on all aspects of our lives and society. Here is a flavour.

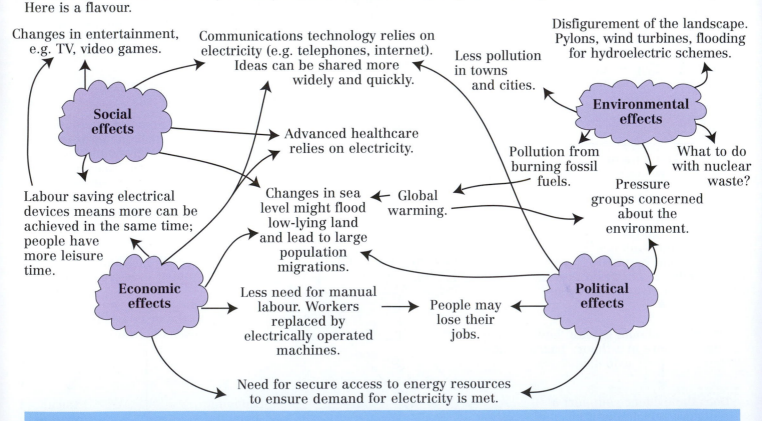

Questions
1. One hundred years ago open coal fires heated many homes. Now electric heaters heat many houses. Suggest some reasons why.
2. Draw an energy flow diagram for three more electrical devices found in your home not shown above.
3. Make a list, with justification of each, of one positive and one negative impact of electricity socially, politically, environmentally and economically not shown above.

THE SUPPLY AND USE OF ELECTRICAL ENERGY
What Influences the Energy Resources We Use?

Electricity provides the majority of the energy needs of the UK. The demand for electricity is predicted to continue to rise. Electricity is a secondary energy source; another (primary) energy source is needed to generate it.

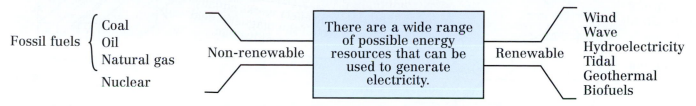

Fossil fuels { Coal, Oil, Natural gas } Nuclear — Non-renewable

There are a wide range of possible energy resources that can be used to generate electricity.

Renewable — Wind, Wave, Hydroelectricity, Tidal, Geothermal, Biofuels

The following pages outline how electricity is generated from these resources, but how do we decide what resources to use? There are a huge number of questions to be answered, and many of the answers may be contradictory.

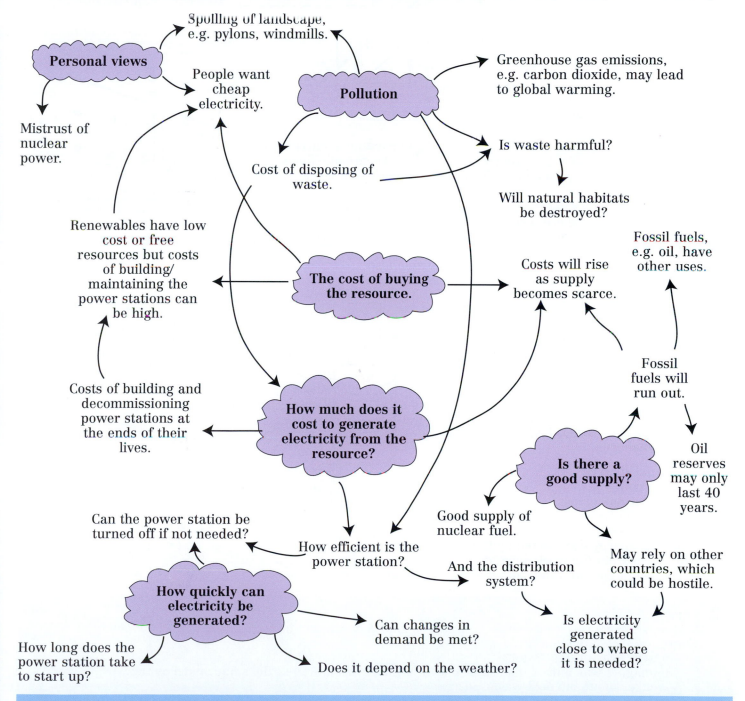

Personal views

Mistrust of nuclear power.

Spoiling of landscape, e.g. pylons, windmills.

People want cheap electricity.

Pollution

Greenhouse gas emissions, e.g. carbon dioxide, may lead to global warming.

Is waste harmful?

Cost of disposing of waste.

Will natural habitats be destroyed?

Renewables have low cost or free resources but costs of building/maintaining the power stations can be high.

The cost of buying the resource.

Fossil fuels, e.g. oil, have other uses.

Costs will rise as supply becomes scarce.

Costs of building and decommissioning power stations at the ends of their lives.

How much does it cost to generate electricity from the resource?

Fossil fuels will run out.

Is there a good supply?

Oil reserves may only last 40 years.

Good supply of nuclear fuel.

Can the power station be turned off if not needed?

How efficient is the power station?

And the distribution system?

May rely on other countries, which could be hostile.

How quickly can electricity be generated?

Can changes in demand be met?

Is electricity generated close to where it is needed?

How long does the power station take to start up?

Does it depend on the weather?

Electricity Generation – Electromagnetic Induction

Moving an electrical conductor through a magnetic field produces a potential difference across the conductor. This is called *electromagnetic induction*.

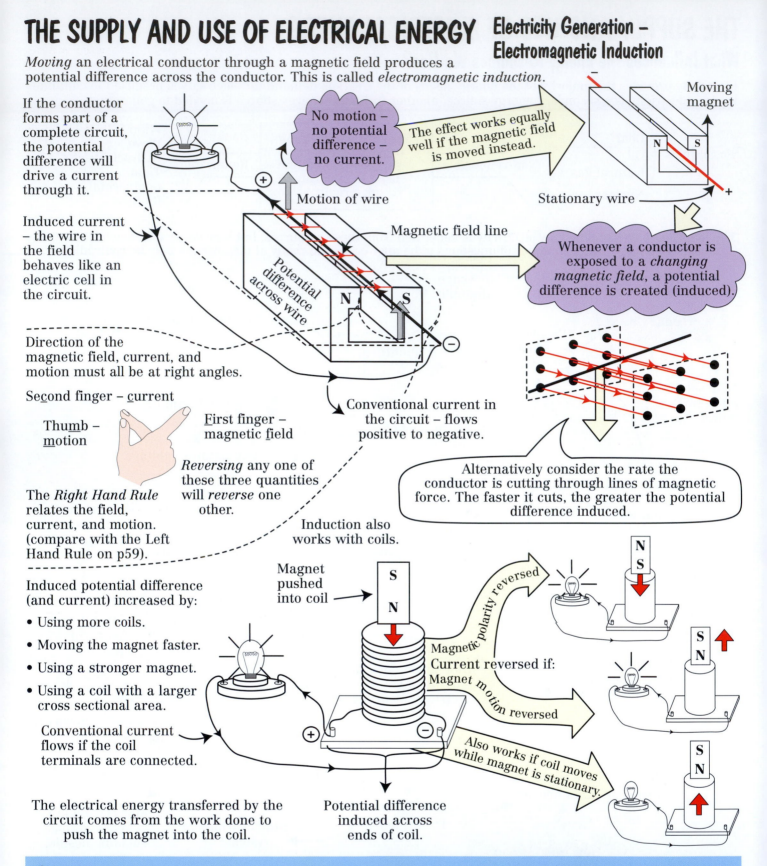

If the conductor forms part of a complete circuit, the potential difference will drive a current through it.

Induced current – the wire in the field behaves like an electric cell in the circuit.

No motion – no potential difference – no current.

The effect works equally well if the magnetic field is moved instead.

Moving magnet

Motion of wire

Magnetic field line

Stationary wire

Whenever a conductor is exposed to a *changing magnetic field*, a potential difference is created (induced).

Potential difference across wire

N   S

Direction of the magnetic field, current, and motion must all be at right angles.

Second finger – <u>c</u>urrent

Thumb – <u>m</u>otion

First finger – magnetic <u>f</u>ield

Reversing any one of these three quantities will *reverse* one other.

The *Right Hand Rule* relates the field, current, and motion. (compare with the Left Hand Rule on p59).

Conventional current in the circuit – flows positive to negative.

Alternatively consider the rate the conductor is cutting through lines of magnetic force. The faster it cuts, the greater the potential difference induced.

Induction also works with coils.

Induced potential difference (and current) increased by:

• Using more coils.

• Moving the magnet faster.

• Using a stronger magnet.

• Using a coil with a larger cross sectional area.

Conventional current flows if the coil terminals are connected.

Magnet pushed into coil

Magnetic polarity reversed

Current reversed if:
Magnet motion reversed

Also works if coil moves while magnet is stationary.

The electrical energy transferred by the circuit comes from the work done to push the magnet into the coil.

Potential difference induced across ends of coil.

Questions
1. A wire is moved at right angles to a magnetic field. What would happen to the size of the potential difference across the wire if:
   a. The wire was moved faster?
   b. The magnet was moved instead of the wire, but it was moved at the same speed as the wire?
   c. A weaker magnetic field was used?
   d. The wire stopped moving?
   e. Two magnets were used end to end so more wire was in the field?
   f. The wire moved from a north pole to a south pole along the magnetic field lines?
2. When pushing a magnet into a coil how could you make the size of the induced potential difference bigger (3 ways)? How could you reverse the direction of the potential difference (2 ways)?
3. When generating electricity by induction where does the energy that is converted into electrical energy come from?

These effects can be used to build a generator.

Contact brushes – provide a sliding contact to allow the coil to rotate and allow the current to flow to an external circuit.

Current direction predicted by the Right Hand Rule

Current   Motion   Field

The output potential difference can be increased by:

Increasing speed of rotation.

Increasing the number of turns on the coil.

Increasing strength of magnetic field.

Placing an iron core in the coil.

Coil

Making the area of the coil larger.

Wire is moving in the opposite direction on the other side of the coil. Therefore current is in the opposite direction so current flows one way around the coil.

Slip rings

Magnets

Output is alternating as the current reverses once every half revolution.

Generators for mains electricity work in a similar way – the output is alternating current.

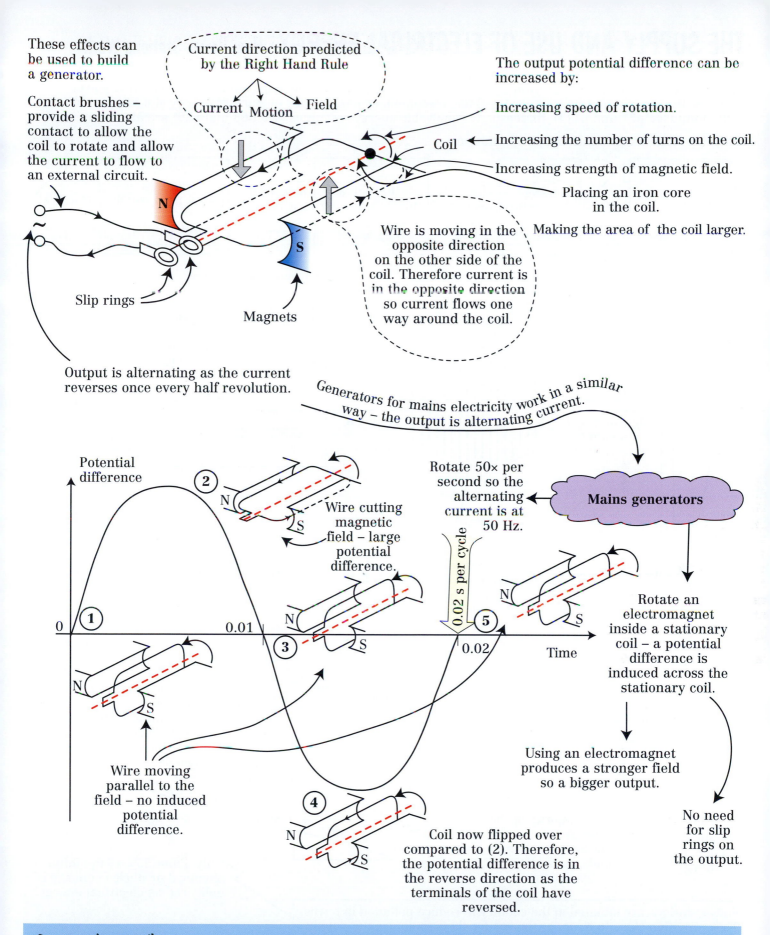

Potential difference

Wire cutting magnetic field – large potential difference.

Rotate 50× per second so the alternating current is at 50 Hz.

0.02 s per cycle

Mains generators

Wire moving parallel to the field – no induced potential difference.

Rotate an electromagnet inside a stationary coil – a potential difference is induced across the stationary coil.

Using an electromagnet produces a stronger field so a bigger output.

Coil now flipped over compared to (2). Therefore, the potential difference is in the reverse direction as the terminals of the coil have reversed.

No need for slip rings on the output.

Questions (continued)
4. List five ways the output of an alternating current generator can be increased.
5. The mains electricity in the UK alternates through 50 complete cycles per second. What does this tell us about the rate of rotation of the generators in power stations in the UK?
6. Suggest two differences between the simple generator shown above and the generators used to generate mains electricity.
7. Why is the potential difference produced by a generator zero twice every revolution?
8. Draw a labelled diagram of an alternating current generator and use it to explain why the current it produces is alternating.

THE SUPPLY AND USE OF ELECTRICAL ENERGY How Power Stations Work

Electricity is very useful energy source because it is easy to distribute and control. However, it is a *secondary energy source* because another primary energy source has to be used to generate it. In conventional power stations, that energy source is either fossil fuels (coal, oil. or natural gas) or nuclear energy stored in uranium or plutonium (see p77 and p111). Increasingly renewable energy resources (see p88 and p89) are also being used.

Here we focus on conventional power stations.

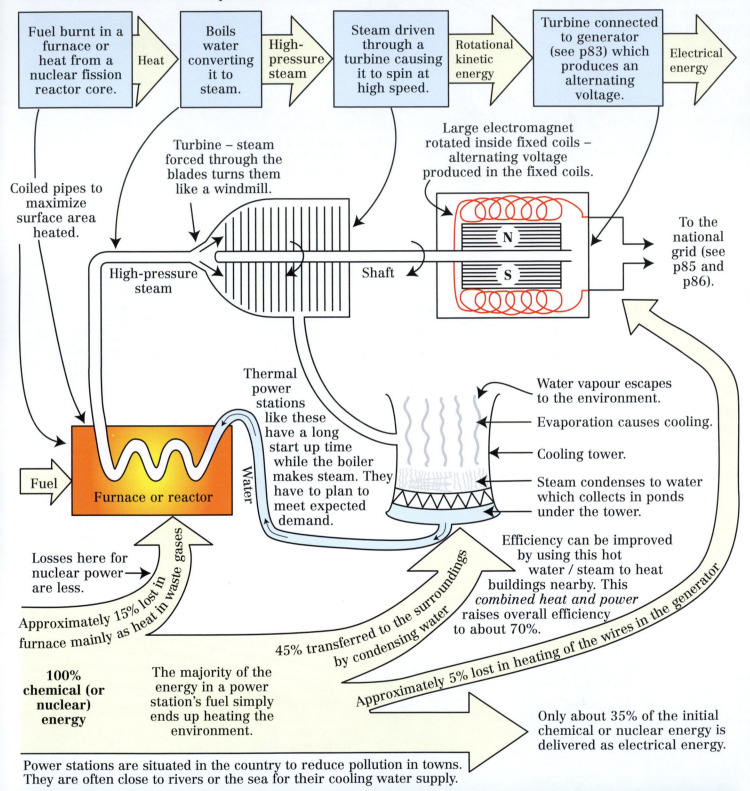

Fuel burnt in a furnace or heat from a nuclear fission reactor core.

Heat

Boils water converting it to steam.

High-pressure steam

Steam driven through a turbine causing it to spin at high speed.

Rotational kinetic energy

Turbine connected to generator (see p83) which produces an alternating voltage.

Electrical energy

Coiled pipes to maximize surface area heated.

Turbine – steam forced through the blades turns them like a windmill.

Large electromagnet rotated inside fixed coils – alternating voltage produced in the fixed coils.

High-pressure steam

Shaft

To the national grid (see p85 and p86).

Thermal power stations like these have a long start up time while the boiler makes steam. They have to plan to meet expected demand.

Water

Fuel

Furnace or reactor

Water vapour escapes to the environment.

Evaporation causes cooling.

Cooling tower.

Steam condenses to water which collects in ponds under the tower.

Losses here for nuclear power are less.

Approximately 15% lost in furnace mainly as heat in waste gases

45% transferred to the surroundings by condensing water

Efficiency can be improved by using this hot water / steam to heat buildings nearby. This *combined heat and power* raises overall efficiency to about 70%.

Approximately 5% lost in heating of the wires in the generator

100% chemical (or nuclear) energy

The majority of the energy in a power station's fuel simply ends up heating the environment.

Only about 35% of the initial chemical or nuclear energy is delivered as electrical energy.

Power stations are situated in the country to reduce pollution in towns. They are often close to rivers or the sea for their cooling water supply.

Questions
1. Name energy sources used to generate electricity in thermal power stations.
2. Draw an energy flow diagram for a coal-fired power station. Start with chemical energy in the coal and end with electrical energy produced.
3. What is combined heat and power?
4. Why are thermal power stations built near rivers or the sea?
5. What is the typical efficiency of conversion of chemical energy to electricity in a thermal power station? To what form of energy is most of the chemical energy converted?

THE SUPPLY AND USE OF ELECTRICAL ENERGY The Transformer

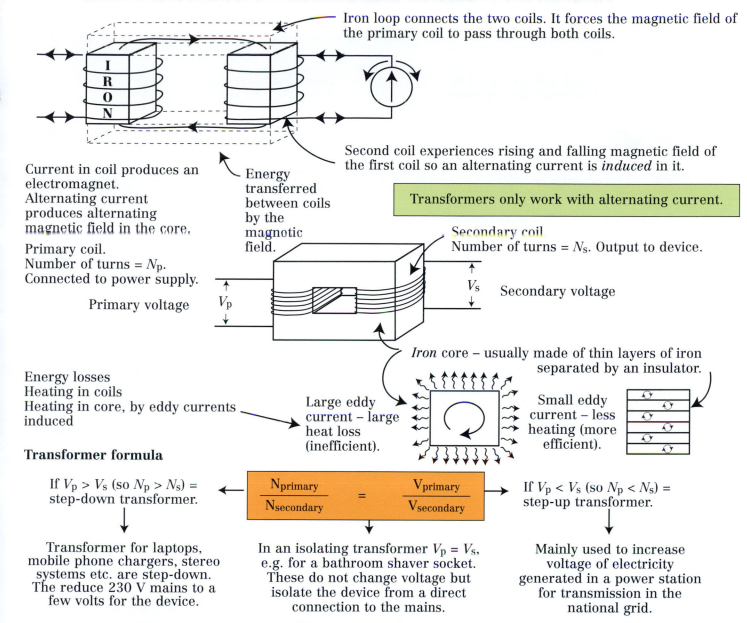

Iron loop connects the two coils. It forces the magnetic field of the primary coil to pass through both coils.

Current in coil produces an electromagnet. Alternating current produces alternating magnetic field in the core.

Primary coil. Number of turns = N_p. Connected to power supply.

Primary voltage

Energy transferred between coils by the magnetic field.

Second coil experiences rising and falling magnetic field of the first coil so an alternating current is *induced* in it.

Transformers only work with alternating current.

Secondary coil. Number of turns = N_s. Output to device.

V_s Secondary voltage

V_p

Iron core – usually made of thin layers of iron separated by an insulator.

Energy losses
Heating in coils
Heating in core, by eddy currents induced

Large eddy current – large heat loss (inefficient).

Small eddy current – less heating (more efficient).

Transformer formula

$$\frac{N_{primary}}{N_{secondary}} = \frac{V_{primary}}{V_{secondary}}$$

If $V_p > V_s$ (so $N_p > N_s$) = step-down transformer.

Transformer for laptops, mobile phone chargers, stereo systems etc. are step-down. The reduce 230 V mains to a few volts for the device.

In an isolating transformer $V_p = V_s$, e.g. for a bathroom shaver socket. These do not change voltage but isolate the device from a direct connection to the mains.

If $V_p < V_s$ (so $N_p < N_s$) = step-up transformer.

Mainly used to increase voltage of electricity generated in a power station for transmission in the national grid.

Questions

1. Copy and complete the following table, giving answers to the nearest whole number:

Transformer	Primary turns	Secondary turns	Primary voltage	Secondary voltage
A		120	240	12
B	625	10 000		400 000
C	20 000		11 000	240
D	2180	1000	240	

Which transformers are step-up and which are step-down?
2. Explain why a transformer needs AC not DC current to work.
3. Remember that electrical power = current × voltage.
   a. If a transformer is supplied with 0.2 A at 240 V, what is the input power?
   b. Assuming the transformer is 100% efficient, what is the output power?
   c. If the ratio $N_p:N_s$ = 2400:60 what is the output voltage?
   d. Hence, what current can be drawn from it?
4. Many people leave mobile phone chargers, containing transformers, plugged in when not in use. The primary coil is connected to the mains, but no current is drawn from the secondary coil by the phone since it is not connected.
   a. How, and from where, does the charger still waste energy?
   b. Even though the energy wasted is small, why should people be encouraged to unplug chargers when not in use?
5. DC electricity is more useful for many applications, but the mains electricity is supplied as AC. Suggest why.

THE SUPPLY AND USE OF ELECTRICAL ENERGY The National Grid

Electricity is supplied from power stations to consumers by a 'national grid' of interconnected cables and transformers. They allow energy to be sent where it is needed anywhere in the country, and diverted around any faults that develop.

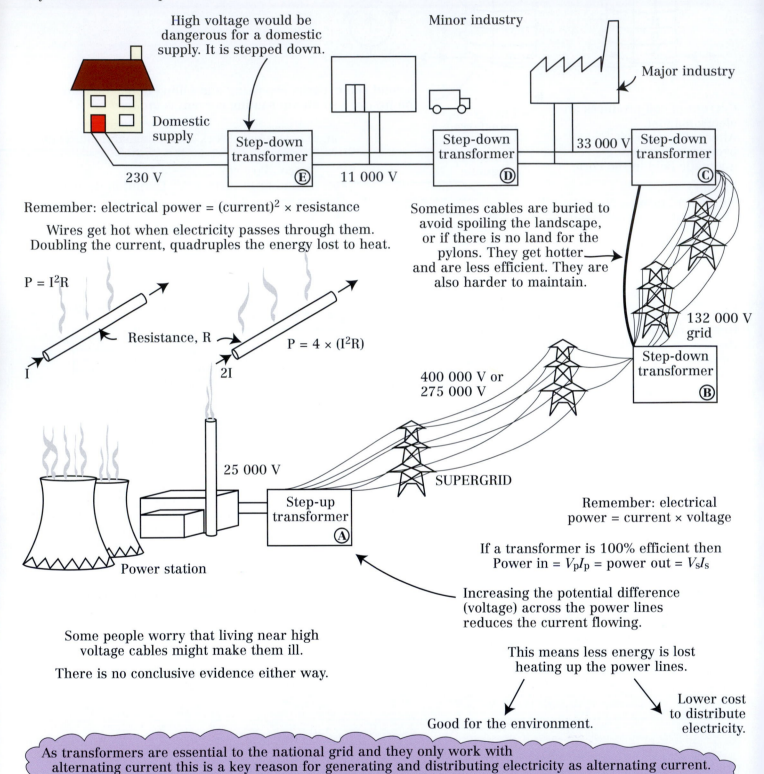

High voltage would be dangerous for a domestic supply. It is stepped down.

Minor industry

Major industry

Domestic supply

Step-down transformer (E)

230 V   11 000 V

Step-down transformer (D)

33 000 V

Step-down transformer (C)

Remember: electrical power = (current)² × resistance

Wires get hot when electricity passes through them. Doubling the current, quadruples the energy lost to heat.

$P = I^2R$

Resistance, R

$P = 4 \times (I^2R)$

I

2I

Sometimes cables are buried to avoid spoiling the landscape, or if there is no land for the pylons. They get hotter and are less efficient. They are also harder to maintain.

132 000 V grid

Step-down transformer (B)

400 000 V or 275 000 V

25 000 V

Step-up transformer (A)

SUPERGRID

Power station

Remember: electrical power = current × voltage

If a transformer is 100% efficient then Power in = V_pI_p = power out = V_sI_s

Increasing the potential difference (voltage) across the power lines reduces the current flowing.

Some people worry that living near high voltage cables might make them ill.

There is no conclusive evidence either way.

This means less energy is lost heating up the power lines.

Good for the environment.

Lower cost to distribute electricity.

As transformers are essential to the national grid and they only work with alternating current this is a key reason for generating and distributing electricity as alternating current.

Questions
1. Suggest two reasons for a 'national grid' to supply electricity, rather than each town having its own power station.
2. Assuming the super-grid power lines operate at 400 000 V, calculate the ratio $N_p:N_s$ for each of the transformers in the diagram above.
3. Why do we use very high voltages to distribute electricity when a lower voltage would be a lot safer?
4. Step-down transformer B (above) has an output of 300 A at 132 000 V, what is the current flowing into it assuming the input voltage is 400 000 V and it is 100% efficient?
5. Explain (using a formula) the statement, 'Doubling the current in a wire, quadruples the energy loss from it as heat'.
6. Draw up a table of advantages and disadvantages of underground vs. overground cables.

THE SUPPLY AND USE OF ELECTRICAL ENERGY

The Environmental Impact of Electricity Generation

All electricity generation has some impact on the environment.

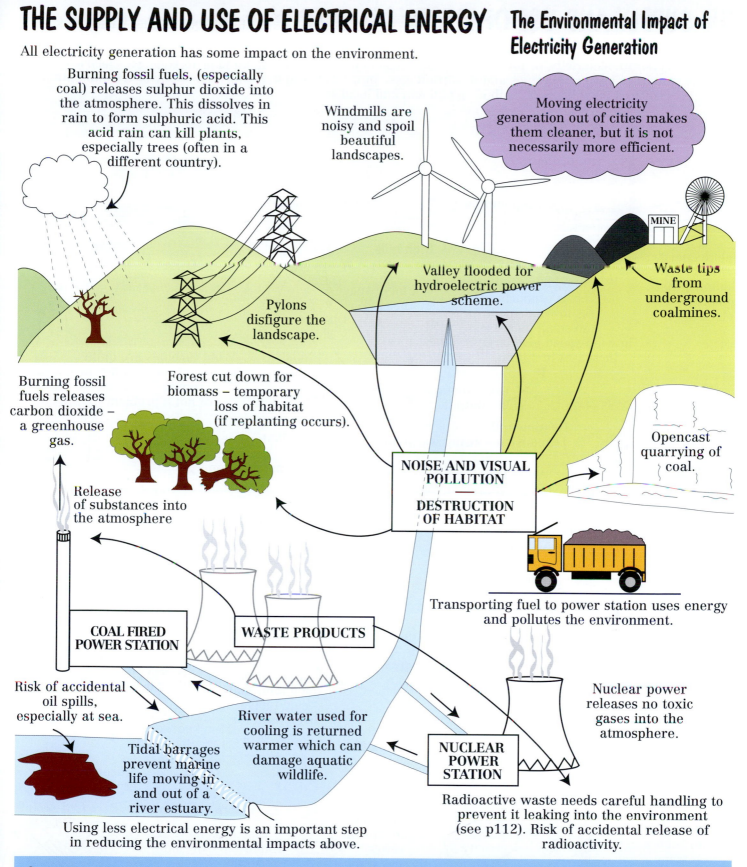

Burning fossil fuels, (especially coal) releases sulphur dioxide into the atmosphere. This dissolves in rain to form sulphuric acid. This acid rain can kill plants, especially trees (often in a different country).

Windmills are noisy and spoil beautiful landscapes.

Moving electricity generation out of cities makes them cleaner, but it is not necessarily more efficient.

MINE

Valley flooded for hydroelectric power scheme.

Waste tips from underground coalmines.

Pylons disfigure the landscape.

Burning fossil fuels releases carbon dioxide – a greenhouse gas.

Forest cut down for biomass – temporary loss of habitat (if replanting occurs).

Release of substances into the atmosphere

NOISE AND VISUAL POLLUTION — DESTRUCTION OF HABITAT

Opencast quarrying of coal.

Transporting fuel to power station uses energy and pollutes the environment.

COAL FIRED POWER STATION

WASTE PRODUCTS

Nuclear power releases no toxic gases into the atmosphere.

Risk of accidental oil spills, especially at sea.

Tidal barrages prevent marine life moving in and out of a river estuary.

River water used for cooling is returned warmer which can damage aquatic wildlife.

NUCLEAR POWER STATION

Radioactive waste needs careful handling to prevent it leaking into the environment (see p112). Risk of accidental release of radioactivity.

Using less electrical energy is an important step in reducing the environmental impacts above.

Questions

1. Make a list of 5 ways you could reduce electricity consumption in your house.
2. In the UK in 2007 there are 1200 wind turbines producing a total of 772 megawatts.
   a. On average how many megawatts does one turbine produce?
   b. All the fossil fuel power stations in the UK combined produce about 60 000 megawatts. How many turbines would be needed to replace all the fossil fuel power stations?
   c. Solar cells produce free electricity without any pollution. Suggest some reasons why they are not very widely used in Britain.
3. The environmental impact of electricity generation is an international problem. Give three examples from above where the impact on the environment could affect more than just the country generating the electricity.
4. Some people say the destruction of a wildlife habitat to build a new dam is not justified. If the dam replaced a coal-fired power station do you agree or not? Justify your argument.

THE SUPPLY AND USE OF ELECTRICAL ENERGY Renewable Energy Resources

Renewable energy resources are those that are *not used up* like fossil fuels. They can be used on a large scale, mainly to generate electricity, or for individual buildings either to provide heating or to generate electricity. All of these resources have advantages and disadvantages. To use renewable resources effectively a combination of different resources must be used, both on a national and local scale.

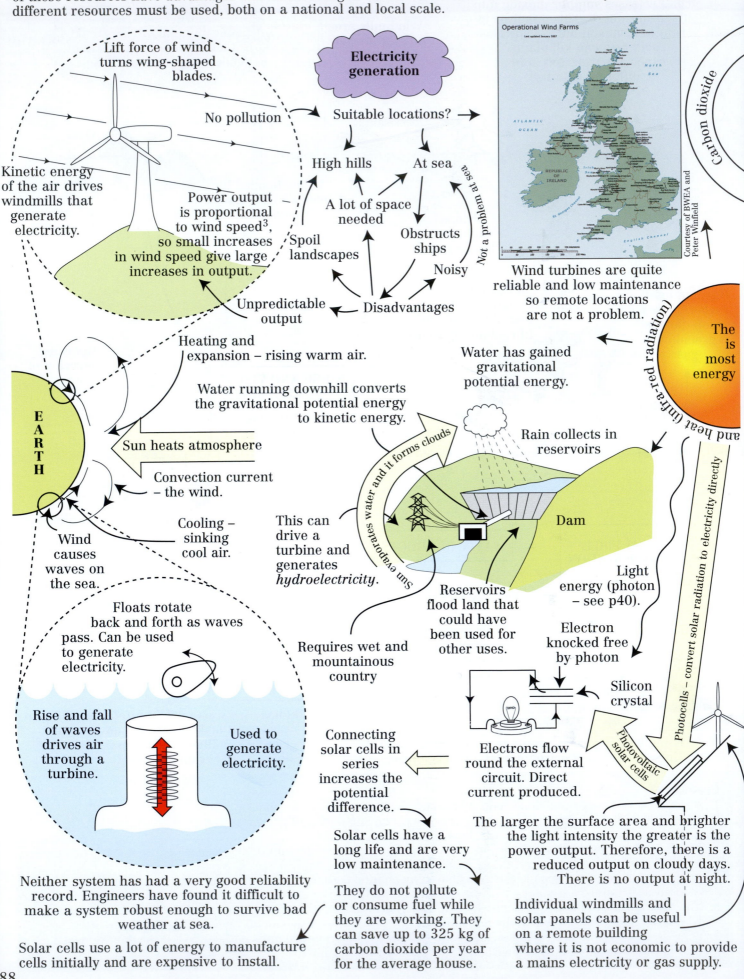

Lift force of wind turns wing-shaped blades.

No pollution

Kinetic energy of the air drives windmills that generate electricity.

Power output is proportional to wind speed³, so small increases in wind speed give large increases in output.

Electricity generation

Suitable locations?

High hills   At sea

A lot of space needed

Spoil landscapes   Obstructs ships

Noisy

Not a problem at sea

Unpredictable output   Disadvantages

Carbon dioxide

Operational Wind Farms

Courtesy of BWEA and Peter Winfield

Wind turbines are quite reliable and low maintenance so remote locations are not a problem.

Heating and expansion – rising warm air.

Water running downhill converts the gravitational potential energy to kinetic energy.

Water has gained gravitational potential energy.

The is most energy

Sun heats atmosphere

EARTH

Convection current – the wind.

Cooling – sinking cool air.

Wind causes waves on the sea.

Sun evaporates water and it forms clouds

This can drive a turbine and generates *hydroelectricity*.

Rain collects in reservoirs

Dam

Reservoirs flood land that could have been used for other uses.

Requires wet and mountainous country

and heat (infra-red radiation)

Light energy (photon – see p40).

Electron knocked free by photon

Silicon crystal

Photocells – convert solar radiation to electricity directly

Floats rotate back and forth as waves pass. Can be used to generate electricity.

Rise and fall of waves drives air through a turbine.

Used to generate electricity.

Electrons flow round the external circuit. Direct current produced.

Photovoltaic solar cells

Connecting solar cells in series increases the potential difference.

Solar cells have a long life and are very low maintenance.

The larger the surface area and brighter the light intensity the greater is the power output. Therefore, there is a reduced output on cloudy days. There is no output at night.

Neither system has had a very good reliability record. Engineers have found it difficult to make a system robust enough to survive bad weather at sea.

Solar cells use a lot of energy to manufacture cells initially and are expensive to install.

They do not pollute or consume fuel while they are working. They can save up to 325 kg of carbon dioxide per year for the average house.

Individual windmills and solar panels can be useful on a remote building where it is not economic to provide a mains electricity or gas supply.

88

The carbon dioxide produced is absorbed by new plants grown to replace those harvested. Overall, it is a carbon neutral process.

Methane burnt as a fuel.

Plants rotting give off methane gas (natural gas).

Large south facing windows provide passive solar heating.

Plants burnt as a fuel (e.g. a wood fire).

Warmed contents of the building radiate long wavelength infrared that cannot pass through the glass.

Heating and alternative fuels

Water pipe

Some systems used curved mirrors to focus the Sun's heat energy onto a pipe containing water, whatever its position in the sky. Solar collectors must face the Sun; south in the northern hemisphere. Ideally, they should move to track the sun.

Solar heating supplements rather than replaces existing water and space heating. It depends on how sunny it is but can reduce domestic carbon dioxide emissions by 400 kg per year.

Biomass

Energy trapped by plants during photosynthesis

The greenhouse effect

Short wavelength infrared

Glass

Energy transferred to Earth as light

Sun our important resource.

Glass case uses this effect

Sunlight falls on solar panels

Water runs through a pipe in the panel.

Hot water

Black background

Cold water

Coiled pipe to maximize surface area.

Hot

Cold

Storage tank

Geothermal heating of homes.

Stored for hot water or pumped through radiators to heat the building.

Electricity generation

Steam

Tides are driven by the Moon, so are very predictable.

Sea – high tide.

At low tide water flows in the opposite direction turning the turbine again.

Dam across river estuary.

Dam can upset the ecosystem and obstructs shipping.

Water flowing in drives turbine to generate electricity.

River – lower level.

Naturally occurring water or water pumped through cracks in the ground.

In volcanic areas, hot rocks are close to surface.

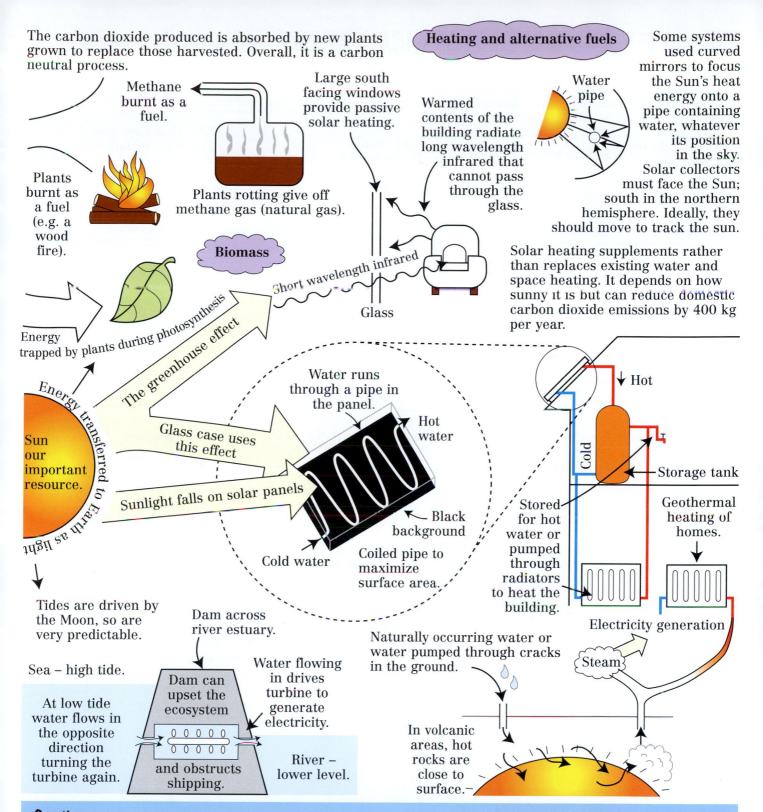

Questions

1. What is our most significant source of energy on Earth?
2. Look at the map of wind farms in the UK. The most common wind direction in the UK is from the southwest. Scotland, Wales and northern England are hilly. Hence, explain why there are very few wind farms in southeast England.
3. Summarize all the information in this chapter in a table as shown.

Source of renewable energy	How it works (you might include a diagram here)	Advantages	Disadvantages or problems

You should be able to include at least eight separate rows.

4. For each of the following, why might people object to having them built near their local area? How might you persuade them to accept the proposal?
   a. A wind farm of twenty large windmills.
   b. A hydroelectric power scheme involving flooding a valley by building a dam across it.
   c. Building a barrage across a river estuary to generate tidal power.
5. UK receives 40% of Europe's total available wind energy but only generates 0.5% of its power from it. Discuss some of the possible reasons why.

Calculating the Cost of the Electrical Energy We Use

Energy is the ability to do *work*. Electrical energy is a convenient way to distribute it to houses, shops, schools, offices, and factories. Like any product, electricity has a *cost* and the more you use the more you pay. Therefore, the amount used has to be measured.

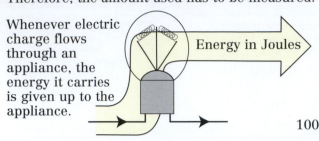

Whenever electric charge flows through an appliance, the energy it carries is given up to the appliance.

Energy in Joules

The number of Joules used by an appliance every second = power.

1 Joule every second = 1 Watt.

1000 Joules per second = 1 kilowatt, kW.

Cost depends on

How long it is switched on for.

Power rating – how quickly the appliance converts electrical energy.

Energy (J) = power (W) × time (s).

Electrical power (W) = current (A) × voltage (V).

Consider a typical device such as a kettle. 4.35 A flows when it is connected to the 230 V mains supply.

Electrical power = 4.35 A × 230 V = 1000 W = 1 kW.

Switched on for 2 mins to boil some water.

Electrical energy = 1000 W × 2 mins × 60 s = 120 000 J.

This would give a huge number of Joules on an electricity bill. A more sensible unit is needed.

If a 1 kW appliance is switched on for 1 hour it uses 1000 W × 3600 s = 3.6 million Joules
We call this a kilowatt-hour or kWh.

Electricity companies call this a unit.

Electricity meter reading

Electrical energy (kWh) = power (kW) × time (h)

N.B. Be careful with the units: power must be in kilowatts not watts.

Energy efficiency
Remember Efficiency =

$$\frac{\text{useful energy output}}{\text{total energy input}}$$

Higher efficiency means:
- Cheaper bills.
- Less pollution.

More efficient
A
B
C
D
E
F
G
Less efficient
A

The cost of the electricity will be   **Number of kilowatt-hours × cost per kilowatt-hour**

Some people can buy cheaper electricity during the night (midnight–8am).

Advantages
Can be used to heat water or 'storage' heaters overnight, which give out their heat during the day.

Disadvantages
You need to use over 20% of your electricity at night, to be cost effective as the daytime rate is usually more expensive than normal. The heat in the water or heaters may have dissipated by the evening – just when you need it most.

The motor effect from p59 can be used to make a practical electric motor.

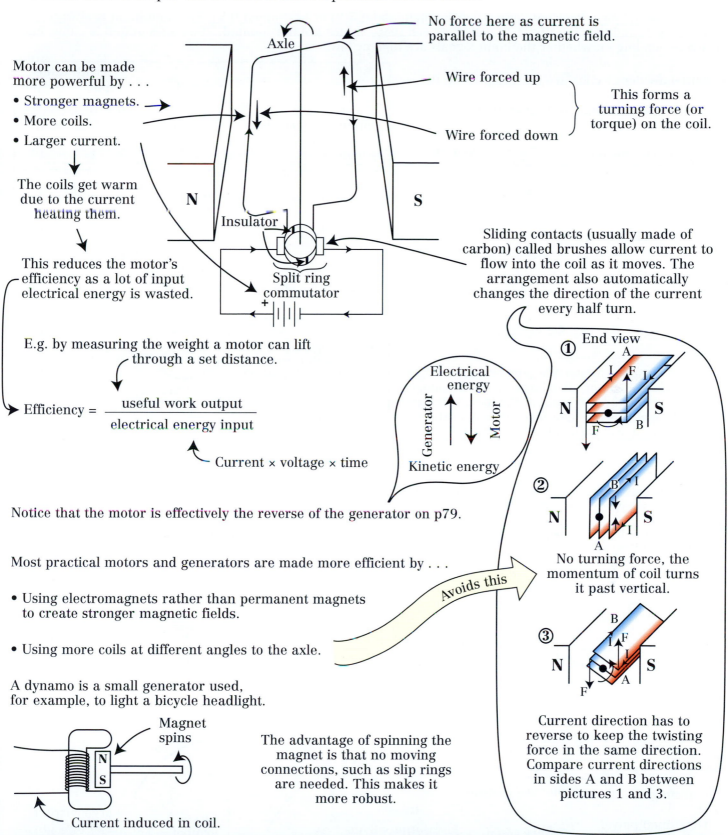

No force here as current is parallel to the magnetic field.

Axle

Wire forced up

This forms a turning force (or torque) on the coil.

Motor can be made more powerful by . . .
• Stronger magnets.
• More coils.
• Larger current.

Wire forced down

The coils get warm due to the current heating them.

N

S

Insulator

Sliding contacts (usually made of carbon) called brushes allow current to flow into the coil as it moves. The arrangement also automatically changes the direction of the current every half turn.

This reduces the motor's efficiency as a lot of input electrical energy is wasted.

Split ring commutator
+

E.g. by measuring the weight a motor can lift through a set distance.

Electrical energy

Generator ↑↓ Motor

Kinetic energy

$$\text{Efficiency} = \frac{\text{useful work output}}{\text{electrical energy input}}$$

Current × voltage × time

① End view
A
N S

Notice that the motor is effectively the reverse of the generator on p79.

Most practical motors and generators are made more efficient by . . .

• Using electromagnets rather than permanent magnets to create stronger magnetic fields.

• Using more coils at different angles to the axle.

Avoids this

②
N S
A

No turning force, the momentum of coil turns it past vertical.

A dynamo is a small generator used, for example, to light a bicycle headlight.

③
N S

Magnet spins

N S

The advantage of spinning the magnet is that no moving connections, such as slip rings are needed. This makes it more robust.

Current direction has to reverse to keep the twisting force in the same direction. Compare current directions in sides A and B between pictures 1 and 3.

Current induced in coil.

Questions
1. How can an electric motor be made more powerful?
2. What would happen to a motor if there was no way of reversing the current direction every half turn? How does a split ring commutator avoid this situation?
3. What are the energy changes in an electric motor? Therefore, why are electric motors not 100% efficient?
4. A motor can lift a weight of 20 N through 3 m in 10 s. If the current flowing is 1.79 A when the voltage of the electricity supply is 12 V, show that the motor is about 30% efficient.
5. What is a dynamo? Explain how it works in as much detail as possible using the ideas from this page and p82–83.

THE SUPPLY AND USE OF ELECTRICAL ENERGY Logic Gates

A logic gate is a circuit that can make decisions depending on the signals it receives.

The input signal for a logic gate can either be high (about 5 V) or low (about 0 V). The high input is always denoted by 1, and low input by 0. Signals between these values are not counted. The gate's output is either high or low depending on whether the input signals are high or low.

Potential divider circuits are used to provide the input voltage for a logic gate that can be either high or low depending on the conditions.

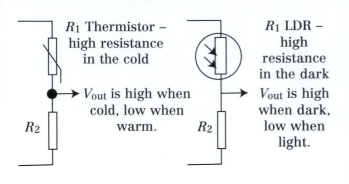

R_1 Thermistor – high resistance in the cold

V_{out} is high when cold, low when warm.

R_1 LDR – high resistance in the dark

V_{out} is high when dark, low when light.

$$V_{out} = R_1 / (R_1+R_2) \times \text{supply p.d.}$$

If R_2 is a variable resistor the temperature or light level at which V_{out} becomes large or small enough to trigger a high or low input at the logic gate can be set. This allows the light level or temperature that the sensor will respond to, to be set.

Name	Symbol	Truth table		
		Input		**Output**
OR Output high (1) when either input (A or B) is high (1)		0	0	0
		1	0	1
		0	1	1
		1	1	1
AND Output high (1) when both inputs (A and B) are high (1)		0	0	0
		1	0	0
		0	1	0
		1	1	1
NOR Output high (1) when neither input (A nor B) is high (1)		0	0	1
		1	0	0
		0	1	0
		1	1	0
NAND Output high (1) unless both inputs (A and B) are high (1)		0	0	1
		1	0	1
		0	1	1
		1	1	0
NOT Reverse the input		0		1
		1		0

Reversing the position of the LDR or thermistor with the fixed resistor, reverses the output.

Logic gates can be used to activate a certain output when required input conditions are met. This can be shown in a *block diagram*.

E.g. A courtesy light switches on in a car when either driver or passenger door, or both, are opened.

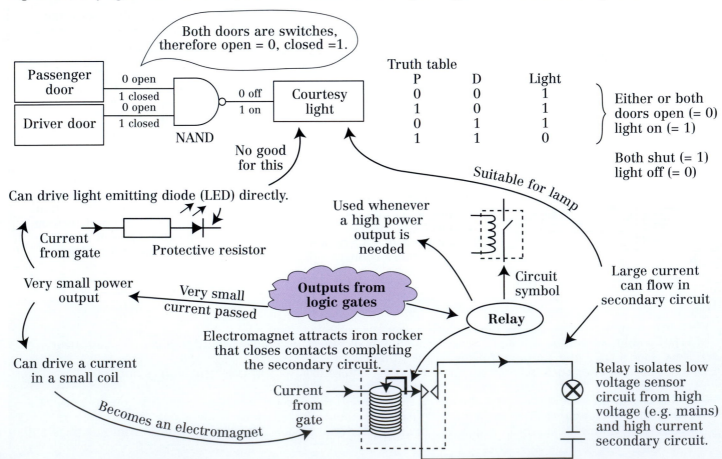

Both doors are switches, therefore open = 0, closed = 1.

Truth table

P	D	Light
0	0	1
1	0	1
0	1	1
1	1	0

Either or both doors open (= 0) light on (= 1)

Both shut (= 1) light off (= 0)

Can drive light emitting diode (LED) directly.

Current from gate — Protective resistor

Very small power output

Can drive a current in a small coil

Becomes an electromagnet

Outputs from logic gates

Very small current passed

Used whenever a high power output is needed

Circuit symbol

Relay

Suitable for lamp

Large current can flow in secondary circuit

Electromagnet attracts iron rocker that closes contacts completing the secondary circuit.

Current from gate

Relay isolates low voltage sensor circuit from high voltage (e.g. mains) and high current secondary circuit.

Designing a logic system:

1. Draw up a block diagram with logic gates

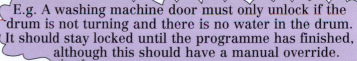

E.g. A washing machine door must only unlock if the drum is not turning and there is no water in the drum. It should stay locked until the programme has finished, although this should have a manual override.

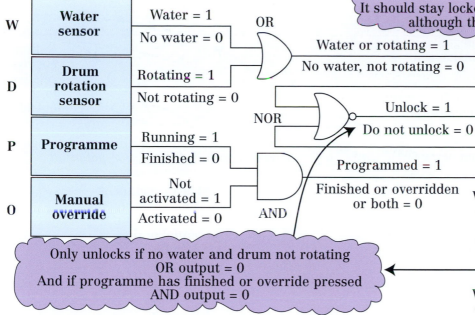

| W | Water sensor | Water = 1 | OR |
| | | No water = 0 | |

Water or rotating = 1
No water, not rotating = 0

| D | Drum rotation sensor | Rotating = 1 |
| | | Not rotating = 0 |

NOR

Unlock = 1
Do not unlock = 0

2. Draw a truth table and check it gives the desired output.

| P | Programme | Running = 1 |
| | | Finished = 0 |

Programmed = 1

| O | Manual override | Not activated = 1 |
| | | Activated = 0 |

Finished or overridden or both = 0

AND

Only unlocks if no water and drum not rotating
OR output = 0
And if programme has finished or override pressed
AND output = 0

W	D	OR output
0	0	0
1	0	1
0	1	1
1	1	1

W	D	AND output
0	0	1
1	0	0
0	1	0
1	1	0

OR output	AND output	NOR output
0	0	1
1	0	0
0	1	0
1	1	0

Latching circuits Sometimes an output is needed which does not change even if the input is removed, e.g. an alarm that continues to ring even if the input that triggered it is removed.

E.g. burglar alarm

| Brief high signal at one input | Set |
| Results in a permanent high signal at the latch output | |

Pressure switch closed by being stood on.

Reset switch

Alarm sounds

| Brief high signal at the other input | Reset |
| Results in a low signal at the latch output | |

Switch opens but

Alarm continues to ring

Alarm stops ringing when reset is closed

A low signal at both inputs does not change the output.
This circuit is called a *bistable* – it has two stable states.

Questions

1. What do the numbers '1' and '0' represent in a truth table?
2. A microwave oven must not start unless the door is closed and the timer is set. Draw a block diagram with a suitable logic gate for this, and include the truth table.
3. In the following circuit, the resistance of the thermistor (R_1) at 100°C is 1.2 kΩ. What resistance should the variable resistor R_2 be set to so V_{out} = 5 V when the temperature reaches 100°C?
4. What is a relay and where are they used? Draw a labelled diagram.
5. In a greenhouse, automatic shades should be drawn if the soil around the plants becomes too dry and if the light level or the temperature rises too much. Draw a suitable block diagram using logic gates and give its truth table.
6. Draw a truth table for the circuit shown:
7. A water pipe may burst if the temperature drops below freezing. Draw a suitable block diagram using two logic gates for a system that will shut off the water to a house if the temperature falls below freezing and not switch it back on until it is reset by a plumber who has inspected the pipes for damage. Draw a truth table for your system.

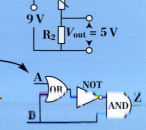

THE SUPPLY AND USE OF ELECTRICAL ENERGY   Electricity and the Human Body

The body sends electrical signals, via nerves from the brain to stimulate muscles.

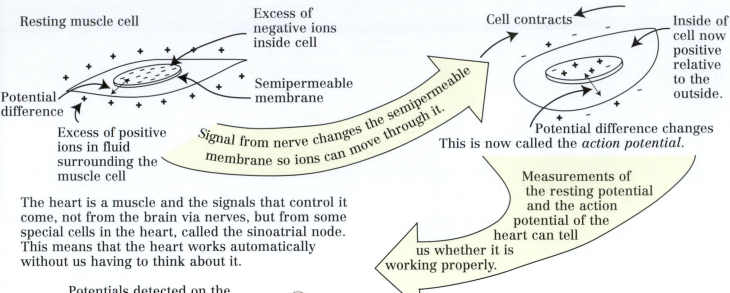

Resting muscle cell

Excess of negative ions inside cell

Semipermeable membrane

Potential difference

Excess of positive ions in fluid surrounding the muscle cell

Signal from nerve changes the semipermeable membrane so ions can move through it.

Cell contracts

Inside of cell now positive relative to the outside.

Potential difference changes This is now called the *action potential*.

Measurements of the resting potential and the action potential of the heart can tell us whether it is working properly.

The heart is a muscle and the signals that control it come, not from the brain via nerves, but from some special cells in the heart, called the sinoatrial node. This means that the heart works automatically without us having to think about it.

Potentials detected on the surface of the body as they are transmitted from the heart by conducting body tissues.

The measurement of these very small potential differences is called an *electrocardiogram* or ECG.

Defibrillation
Fibrillation occurs when a patient's heart does not beat rhythmically but quivers. Blood is not pumped and the patient will soon die.

Paddle electrodes   3000 V

20 A

Current passed for 5 milliseconds

All the heart muscles contract strongly.

Usually when they relax, they settle back into a regular rhythm of beating.

Care must be taken not to shock the operator.

① Blood arrives from the lungs and the rest of the body and collects in the atria.

Blood pushed into the ventricles

Pacemakers

- Cells producing stimulus to the heart stop functioning.
- Electrical devices are fitted producing tiny, but regular, shocks to the heart.
- It is placed under the skin and a wire fed through a vein to the heart.
- Fitted with long-life batteries.

Potential difference at electrodes / mV

1

② Ventricles contract

Atria contract

Blood forced out to lungs and the rest of the body.

0.1 s

Atria relax as ventricles contract and potential difference for this is masked by a large potential difference as the ventricles contract.

③ Ventricles relax

94

TRANSPORT Stopping Distances

Reaction time is the time it takes a driver to see a hazard, recognize they need to take action, decide on the action, and use the vehicle's controls.

E.g. a child runs into the road, the driver sees the child and considers braking or swerving. They decide to brake and press the brake pedal. Normally all these stages happen in about 1 second.

Affected by
- Driver tiredness
- Distractions – for example using a mobile phone
- Influence of alcohol or drugs

Reaction time

Affected by
- Vehicle speed

Braking distance is the distance travelled between the driver first pressing the pedal and the vehicle stopping.

Affected by
- The speed the vehicle was travelling at
- The mass of the vehicle
- Weather conditions – a wet or icy road for example
- The condition of the tyres and brakes (i.e. available brake force)
- Quality of the road surface

Thinking distance – distance travelled during driver's reaction time.

Thinking distance

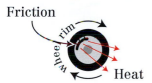

Braking distance

The total stopping distance of the vehicle = thinking distance + braking distance.

When braking, the brakes do work against friction.

Friction

wheel rim

Heat

This converts the kinetic energy of the vehicle into heat energy.

Braking distance proportional to velocity²

Stopping distance

Based on brake force = 6.5 kN and 1 tonne car

Thinking distance = 3 m per 10 mph

Vehicle velocity (mph)

Therefore, a small increase in vehicle speed gives a large increase in braking distance.

Hence kinetic energy = work done against friction

$$\tfrac{1}{2}\ \text{(mass of vehicle)} \times \text{(velocity of vehicle)}^2 = \text{brake force} \times \text{distance travelled while braking.}$$

$$\tfrac{1}{2}\ mv^2 = F \times d.$$

Therefore minimum braking distance $= \tfrac{1}{2}\ mv^2/F$

Notice the braking distance is proportional to the square of the velocity. This is reflected in the stopping distances given in the Highway Code. (Crown copyright)

20 MPH	6 m 6 m		= 12 m or 3 car lengths
30 MPH	9 m	14 m	= 23 m or 5 car lengths
40 MPH	12 m	24 m	= 36 m or 9 car lengths
50 MPH	15 m	38 m	= 53 m or 13 car lengths
60 MPH	18 m	55 m	= 73 m or 18 car lengths
70 MPH	21 m	75 m	= 96 m or 24 car lengths

■ Thinking distance
■ Breaking distance
Average car length = 4 metres

Braking distance is proportional to the mass of the vehicle assuming the same brake force. Therefore, large vehicles need brakes that can exert a larger force.

Questions
1. Write a list of factors that affect braking distance.
2. As the speed of a vehicle increases what happens to the size of the brake force needed to stop it in a certain distance?
3. Explain why each of the following is a driving offence that the police might stop you for in terms of their effect on the thinking distance or braking distance of a vehicle. The first one has been done as an example:
   a. Driving faster than the speed limit. Answer – increases both thinking distance and braking distance so a vehicle is less likely to be able to stop in the distance the driver can see to be clear.
   b. Having bald tyres. c. Driving under the influence of alcohol. d. Using a mobile phone while driving.
4. Copy and complete the following table:

Mass of vehicle (kg)	Initial velocity (m/s)	Maximum braking force (N)	Minimum braking distance (m)
1000	13.3 (= 30 mph)	6500	
1000	31.1 (= 70 mph)	6500	
5000	13.3 (= 30 mph)	6500	

Explain why the braking distances you calculated are minimum distances.

TRANSPORT Road Safety

Most car safety features are designed to reduce the force of any collision on the passengers, which reduce the injuries they may suffer.

Velocity, v

Momentum = mv

Velocity = 0

Car decelerates over time = t

Force, F

Momentum = 0

Change in momentum = mv

Deceleration = change in velocity / time taken

$$= (v - 0)/t$$

Therefore as force = mass × deceleration

$$= m(v - 0)/t$$

$$F = mv/t$$

Force (N) × time (s) = change of momentum (kgm/s) (see p16)

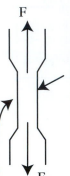

Seatbelts are designed to extend during a collision, making the collision last longer and reducing the force. They also restrain the passengers, preventing them hitting hard surfaces such as the dashboard or other passengers.

Once a seatbelt has been in a crash it has been permanently stretched, by absorbing some of the kinetic energy of the passenger wearing it. It will not absorb kinetic energy as effectively again so has to be replaced.

Antilock brakes (ABS) automatically, briefly, release the brakes if the wheels stop turning (lock) during braking. This helps to prevent skidding.

A crumple zone is a specially weakened part of the vehicle's chassis. This deforms in a collision making the collision last longer and reducing the forces exerted on the passengers.

All of these features also absorb kinetic energy by deforming (changing shape).

Crash barriers beside roads prevent vehicles crossing into the path of oncoming traffic.

They are designed to deform on collision absorbing the kinetic energy of the vehicle.

An airbag is designed to make the passenger stop moving more gradually, increasing the time of collision. This reduces the force.

Airbag is triggered by an acceleration sensor. A very large deceleration, such as in a collision triggers a small explosive device, producing a lot of gas very quickly. This inflates the bag. The gas escapes from small holes in the back of the bag as the passenger sinks into it, cushioning them.

From p95, braking distance is inversely proportional to brake force.

Small force, just due to friction with the road

Friction

Locked wheel

Large force due to friction with road *and* with brakes as wheels turn.

Friction

Rotating wheel

Skid – large stopping distance.

(see p16)

Questions
1. Using the words 'force', 'deceleration', 'momentum', 'energy' in your answers explain:
   a. How motorcycle and cycle helmets made of a material that deforms on impact with a hard surface, reduce the severity of head injuries to the rider in a collision.
   b. Why crash barriers alongside roads are often made of deformable material like steel tube rather than a rigid material like concrete.
   c. An escape lane (a pit full of deep sand) is provided at the bottom of steep hills for drivers to steer into if their brakes fail.
2. A car passenger of mass 70 kg is travelling at 13.3 m/s (30 mph). Show that their momentum is 931 kgm/s. In a collision, they hit the dashboard and stop in 0.01 s. Show that the force exerted is about 93 kN. In another collision, the passenger is cushioned by an air bag and stops in 0.1 s. Show the force is now only 9.3 kN.
3. A car of mass 1000 kg travelling at 13.3 m/s hits a brick wall and stops in 0.1 s. Calculate the deceleration. What force is exerted on the car by the wall? The same car is now fitted with a crumple zone and stops in 0.5 s. What force is exerted by the wall now?
4. Explain how antilock brakes (ABS) can help to reduce stopping distances when a driver brakes hard.

WAVES AND COMMUNICATIONS Using Waves to Communicate

The long distance transmission of information, other than by a message on paper or via an electrical signal in a wire, relies on using waves.

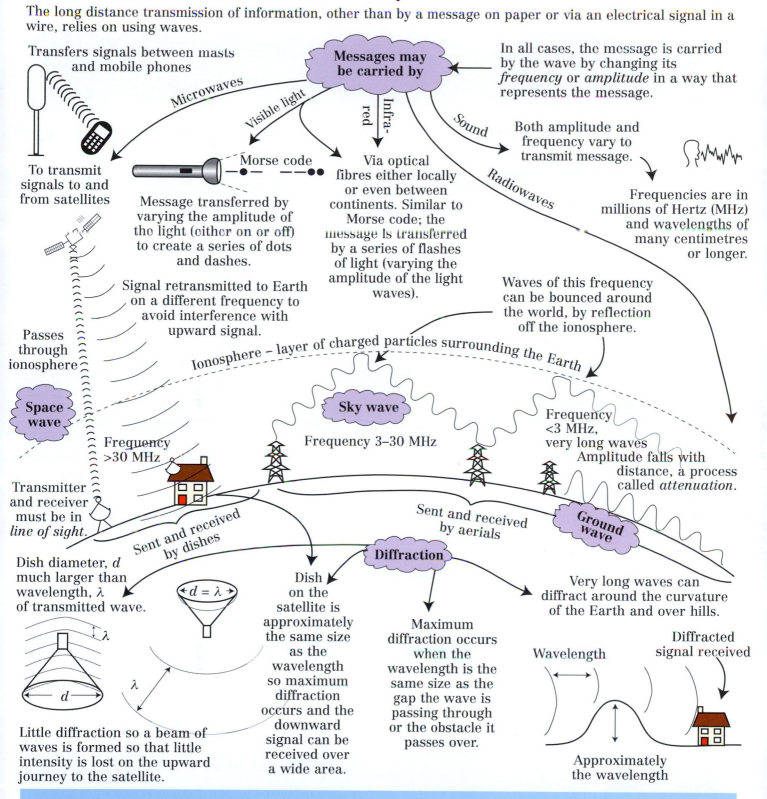

Transfers signals between masts and mobile phones

Messages may be carried by

In all cases, the message is carried by the wave by changing its *frequency* or *amplitude* in a way that represents the message.

Microwaves

Visible light

Infra-red

Sound

Radiowaves

To transmit signals to and from satellites

Morse code

Message transferred by varying the amplitude of the light (either on or off) to create a series of dots and dashes.

Via optical fibres either locally or even between continents. Similar to Morse code; the message is transferred by a series of flashes of light (varying the amplitude of the light waves).

Both amplitude and frequency vary to transmit message.

Frequencies are in millions of Hertz (MHz) and wavelengths of many centimetres or longer.

Signal retransmitted to Earth on a different frequency to avoid interference with upward signal.

Waves of this frequency can be bounced around the world, by reflection off the ionosphere.

Passes through ionosphere

Ionosphere – layer of charged particles surrounding the Earth

Space wave

Sky wave

Frequency <3 MHz, very long waves Amplitude falls with distance, a process called *attenuation*.

Frequency >30 MHz

Frequency 3–30 MHz

Transmitter and receiver must be in *line of sight*.

Sent and received by dishes

Sent and received by aerials

Ground wave

Diffraction

Dish diameter, d much larger than wavelength, λ of transmitted wave.

$d = \lambda$

Dish on the satellite is approximately the same size as the wavelength so maximum diffraction occurs and the downward signal can be received over a wide area.

Maximum diffraction occurs when the wavelength is the same size as the gap the wave is passing through or the obstacle it passes over.

Very long waves can diffract around the curvature of the Earth and over hills.

Wavelength

Diffracted signal received

Little diffraction so a beam of waves is formed so that little intensity is lost on the upward journey to the satellite.

Approximately the wavelength

Questions
1. Name four types of electromagnetic waves used to send messages. Suggest why the other types of electromagnetic radiation are unsuitable.
2. Explain three ways radiowaves can be used to send messages over long distances. Use diagrams to help your explanation.
3. i. Use the formula wave speed = frequency × wavelength to calculate the wavelength of radiowaves of frequency:
   a. 3 MHz (3×10^6 Hz).   b. 1800 MHz (1.8×10^9 Hz).   c. 30 GHz (3×10^{10} Hz).
   ii. Explain which of the above frequencies would be most useful for:
   a. Diffracting around large obstacles like hills. b. Sending to a satellite using a dish.
   c. Mobile telephone communication.
4. A signal is to be sent from the UK to America across the Atlantic. Explain:
   a. Why a signal sent by a ground wave would be very weak by the time it reached America.
   b. Why the ionosphere is needed if the signal is to be sent by a sky wave.
   c. Why a satellite is needed if the signal is to be sent by a space wave.

WAVES AND COMMUNICATIONS Analogue and Digital Signals

Computers work with binary code, a series high or low voltages representing 1 and 0. They are particularly good at processing digital signals.

Communication signals are either

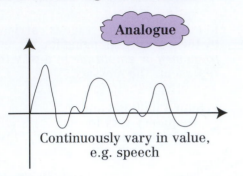

Analogue

Continuously vary in value, e.g. speech

Or

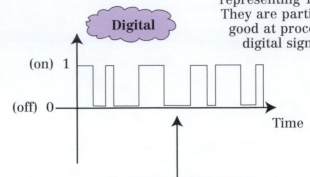

Digital

(on) 1

(off) 0

Time

Two distinct values only. Usually on and off represented by 1 (on) and 0 (off).

Conversion:
Sample too often and the process is too slow.
Sample too few times and not enough information is available to reconstruct the original signal.

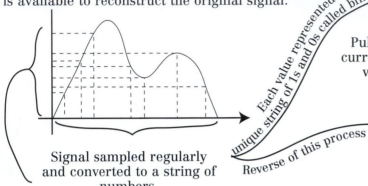

Signal sampled regularly and converted to a string of numbers.

Each value represented by a unique string of 1s and 0s called binary code.

Reverse of this process

Pulses of current in a wire.

Flashes of visible or infrared light in fibre-optic cables.

Pulses of radio or microwaves.

When digital signals are received, they need to be converted back to analogue.

Humans cannot directly interpret digital signals. Our senses respond to analogue signals.

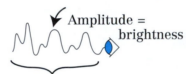

Amplitude = brightness

Frequency = colour

Amplitude = loudness

Frequency = pitch

Advantages of digital signals:

1) Less noise

Noise is any unwanted interference picked up by the signal.

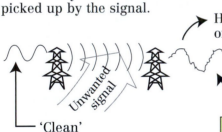

'Clean' original signal

Unwanted signal

Hiss or crackle on sound signal

'Noisy' signal

Distortion of TV picture

Any amplification also amplifies the noise.

With analogue signals it is hard to remove noise.

Digital signals are still clearly 1s and 0s even with noise.

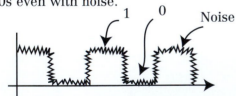

1   0   Noise

When decoded the noise is removed.

Therefore, digital radio and TV have better sound and picture quality.

2) More information can be transmitted at once.

Signals can be interleaved, (called *multiplexing*).

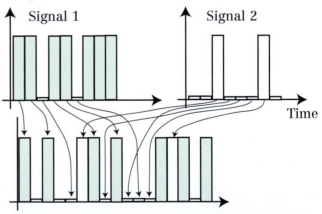

Signal 1

Signal 2

Time

Therefore, lots of information can be sent on one signal. Hence, this is why you can have so many digital TV and radio stations but relatively few analogue ones.

Questions
1. Use diagrams to illustrate the difference between a digital and an analogue signal.
2. When listening to a radio station a hissing sound is heard. What is likely to have caused this and is the signal most likely to have been analogue or digital?
3. Morse code is transmitted as a series of pulses of electricity in a wire or flashes of light representing dots and dashes. Explain whether it is an analogue or digital signal.
4. How are analogue signals converted to digital?
5. What is multiplexing?
6. Explain two advantages of digital signals compared to analogue.
7. When signals are amplified, noise is also amplified. Why is this less of a problem for digital signals?

WAVES AND COMMUNICATIONS AM/FM Radio Transmission

When you tune to a given radio or TV station, you select a particular frequency of radiowave to be received. This wave is called a *carrier wave*, but how is the message added to the carrier wave? There are two methods by which the carrier wave is *modulated* (or varied) by the message signal.

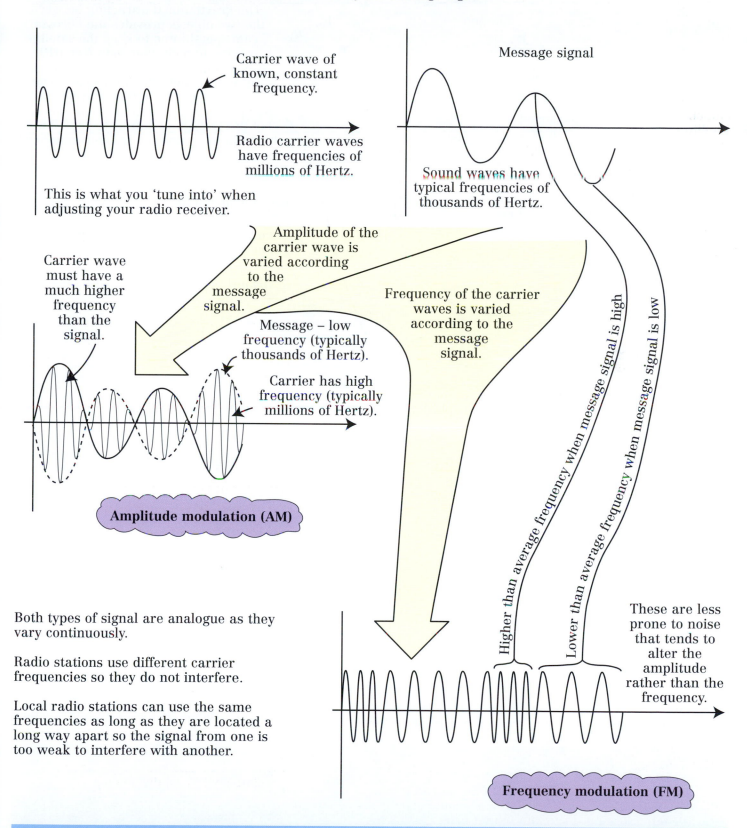

Carrier wave of known, constant frequency.

Radio carrier waves have frequencies of millions of Hertz.

This is what you 'tune into' when adjusting your radio receiver.

Message signal

Sound waves have typical frequencies of thousands of Hertz.

Amplitude of the carrier wave is varied according to the message signal.

Carrier wave must have a much higher frequency than the signal.

Message – low frequency (typically thousands of Hertz).

Carrier has high frequency (typically millions of Hertz).

Frequency of the carrier waves is varied according to the message signal.

Amplitude modulation (AM)

Higher than average frequency when message signal is high

Lower than average frequency when message signal is low

These are less prone to noise that tends to alter the amplitude rather than the frequency.

Both types of signal are analogue as they vary continuously.

Radio stations use different carrier frequencies so they do not interfere.

Local radio stations can use the same frequencies as long as they are located a long way apart so the signal from one is too weak to interfere with another.

Frequency modulation (FM)

Questions
1. What is a carrier wave?
2. What do you understand by the term 'modulation' in the context of radiowaves?
3. What do the abbreviations AM and FM stand for?
4. Use diagrams to explain the difference between AM and FM radio transmissions.
5. Which type of transmission, AM or FM suffers less from noise?
6. Can two different national radio stations covering the whole of the UK use the same carrier wave frequency? What about two local stations?

WAVES AND COMMUNICATIONS   Satellite Orbits and Their Uses

Satellites are objects that orbit larger objects in space. They can be natural, like the moon orbiting the Earth or artificial (man-made).

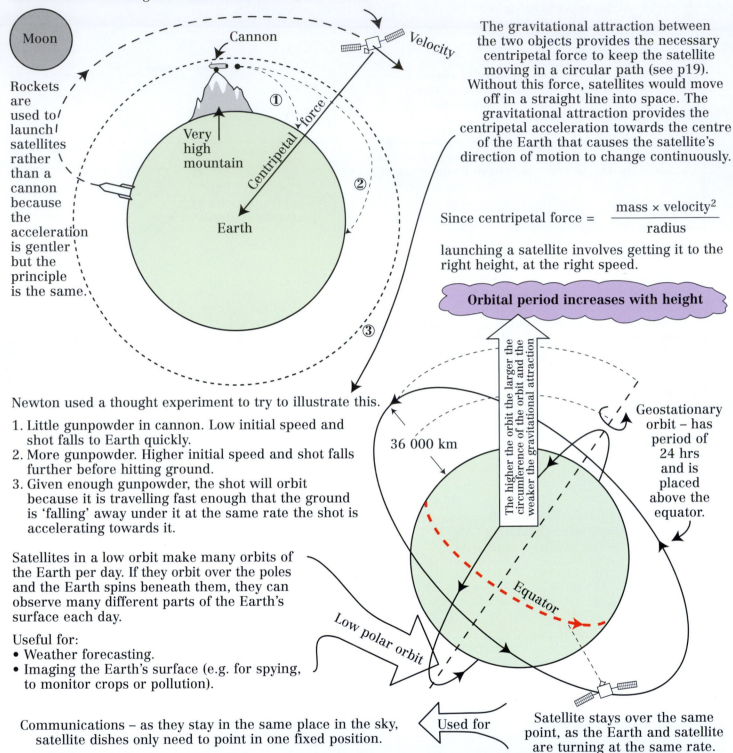

Rockets are used to launch satellites rather than a cannon because the acceleration is gentler but the principle is the same.

The gravitational attraction between the two objects provides the necessary centripetal force to keep the satellite moving in a circular path (see p19). Without this force, satellites would move off in a straight line into space. The gravitational attraction provides the centripetal acceleration towards the centre of the Earth that causes the satellite's direction of motion to change continuously.

Since centripetal force = $\dfrac{\text{mass} \times \text{velocity}^2}{\text{radius}}$

launching a satellite involves getting it to the right height, at the right speed.

Orbital period increases with height

The higher the orbit the larger the circumference of the orbit and the weaker the gravitational attraction

Geostationary orbit – has period of 24 hrs and is placed above the equator.

Newton used a thought experiment to try to illustrate this.

1. Little gunpowder in cannon. Low initial speed and shot falls to Earth quickly.
2. More gunpowder. Higher initial speed and shot falls further before hitting ground.
3. Given enough gunpowder, the shot will orbit because it is travelling fast enough that the ground is 'falling' away under it at the same rate the shot is accelerating towards it.

Satellites in a low orbit make many orbits of the Earth per day. If they orbit over the poles and the Earth spins beneath them, they can observe many different parts of the Earth's surface each day.

Useful for:
• Weather forecasting.
• Imaging the Earth's surface (e.g. for spying, to monitor crops or pollution).

Communications – as they stay in the same place in the sky, satellite dishes only need to point in one fixed position.

Used for

Satellite stays over the same point, as the Earth and satellite are turning at the same rate.

WAVES AND COMMUNICATIONS Images and Ray Diagrams

Light follows straight lines, or rays, from a source of light to an observer unless it is reflected, by a mirror, or refracted, by a lens, on route.

Mirrors and lenses come in a variety of shapes to manipulate the light rays in various useful ways. Ray diagrams help us to understand their effects.

Rays show the direction the light waves are travelling in. Light rays always travel in straight lines (as light waves travel in straight lines) except when reflected or refracted when they change direction.

An image is formed at a point where the light rays from an object appear to come from, had their direction not been changed by a mirror or lens.

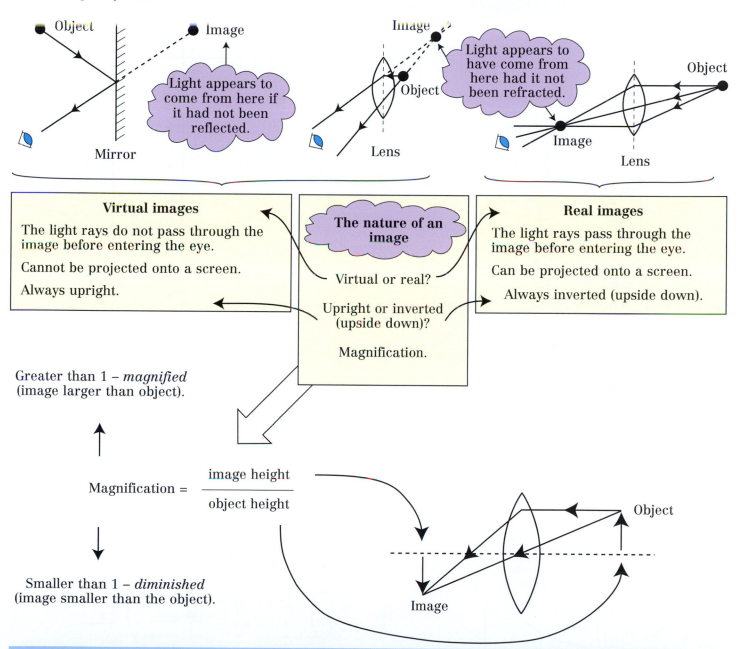

Questions
1. Make a list of three properties of an image that describe the 'nature of an image'.
2. State three differences between a real and virtual image.
3. Is the image in a plane (flat) mirror real or virtual?
4. What is a light ray?
5. What is the formula for magnification? If the magnification of a lens is less than 1, would the image be larger or smaller than the object?
6. A tree has a height of 20 m. In a photograph, it has a height of 20 cm. What is the magnification?
7. A letter 'I' in a book has a height of 5 mm. When viewed through a magnifying glass with a magnification of 1.9, how high will it appear?

101

WAVES AND COMMUNICATIONS Mirrors and Lenses, Images

Mirrors

(1) Plane (flat)

Nature of image
Virtual
Upright
Same size as object

Equal angles

← Silvering

← Normal – a construction line
at right angles to the surface
at the point where a light ray meets it.

Law of reflection (applies to all mirrors):

Angle of incidence, i = angle of reflection, r

Diffuse reflection from a rough
surface – no image formed.

(2) Concave – curving in (like a cave)

Brings light to a focus so is a *converging* mirror.

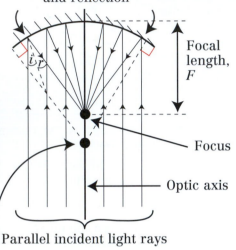

Equal angles of incidence
and reflection

Focal length, F

Focus

Optic axis

Parallel incident light rays

Centre of curvature C – centre
of a sphere that the mirror
forms part of the surface of.

Nature of images

	Object	Image
	Beyond C	Between C and F Real Inverted Diminished
	At C	At C Real Inverted Same size
	Between C and F	Beyond C Real Inverted Magnified
	Closer than F	Virtual Upright Magnified

Rules for drawing ray diagrams for concave mirrors
Ray from the object
1. Parallel to optic axis – reflects through F.
2. To centre of mirror is reflected, forming equal angles with optic axis.
3. Through F is reflected parallel to the optic axis.

(3) Convex – bulges out

Spreads light rays out so is a *diverging* mirror.

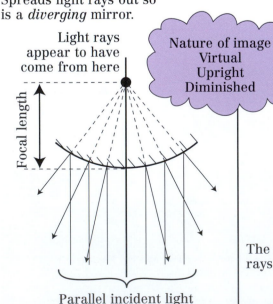

Light rays appear to have come from here

Focal length

Nature of image
Virtual
Upright
Diminished

Parallel incident light rays

Lenses

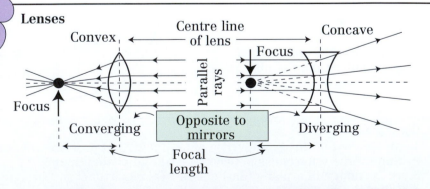

Convex

Centre line of lens

Concave

Focus

Focus

Parallel rays

Converging

Opposite to mirrors

Diverging

Focal length

The more powerful a lens, the greater the change in direction of the light rays, and therefore the closer the focus is to the centre line of the lens.

Power of lens (dioptre) = 1/focal length (m)

The more curved the surface the greater the refraction of the light. Therefore, fat lenses have short focal lengths and are more powerful.

1. Convex (converging) lens

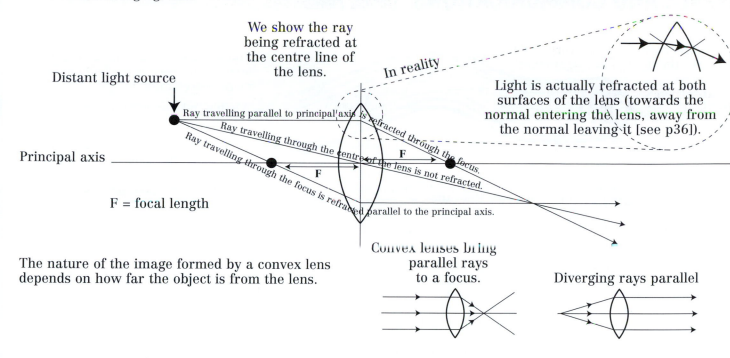

We show the ray being refracted at the centre line of the lens.

Distant light source

Ray travelling parallel to principal axis is refracted through the focus.

In reality

Light is actually refracted at both surfaces of the lens (towards the normal entering the lens, away from the normal leaving it [see p36]).

Ray travelling through the centre of the lens is not refracted.

Principal axis

Ray travelling through the focus is refracted parallel to the principal axis.

F = focal length

The nature of the image formed by a convex lens depends on how far the object is from the lens.

Convex lenses bring parallel rays to a focus.

Diverging rays parallel

	Object	Image	Uses
	Further than 2F	Between F and 2F Real Inverted Diminished	Camera: convex lens focuses light from a distant object to form a diminished image on the film close to the lens
	Between F and 2F	Further than 2F Real Inverted Magnified	Projector: convex lens focuses light from a nearby object to form an enlarged image on a distant screen
	Closer than F	Upright Virtual Magnified	Magnifying glass

2. Concave (diverging) lens

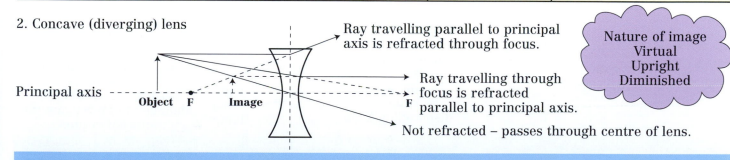

Ray travelling parallel to principal axis is refracted through focus.

Principal axis

Object   F   Image   F

Ray travelling through focus is refracted parallel to principal axis.

Not refracted – passes through centre of lens.

Nature of image
Virtual
Upright
Diminished

Questions
1. Describe what we mean by the term 'focal point'.
2. Draw the shapes of convex and concave mirrors and lenses. Show with ray diagrams which will bring parallel light waves to a focus, and which will diverge them.
3. What three rays are drawn in a ray diagram for:
   a. A convex lens?   b. A concave mirror?
4. Does a powerful lens have a short or long focal length? What unit is the power of a lens measured in?
5. A lens has a focal length of 0.1 m. What is its power?
6. Draw a ray diagram for an object placed at 2F from a convex lens and at F from a convex lens.
7. Draw a ray diagram for a camera and a projector; include the object, image, and lens.

WAVES AND COMMUNICATIONS Optical Fibres

An optical fibre is a thin strand of very clear glass through which visible light or infrared radiation can be guided.

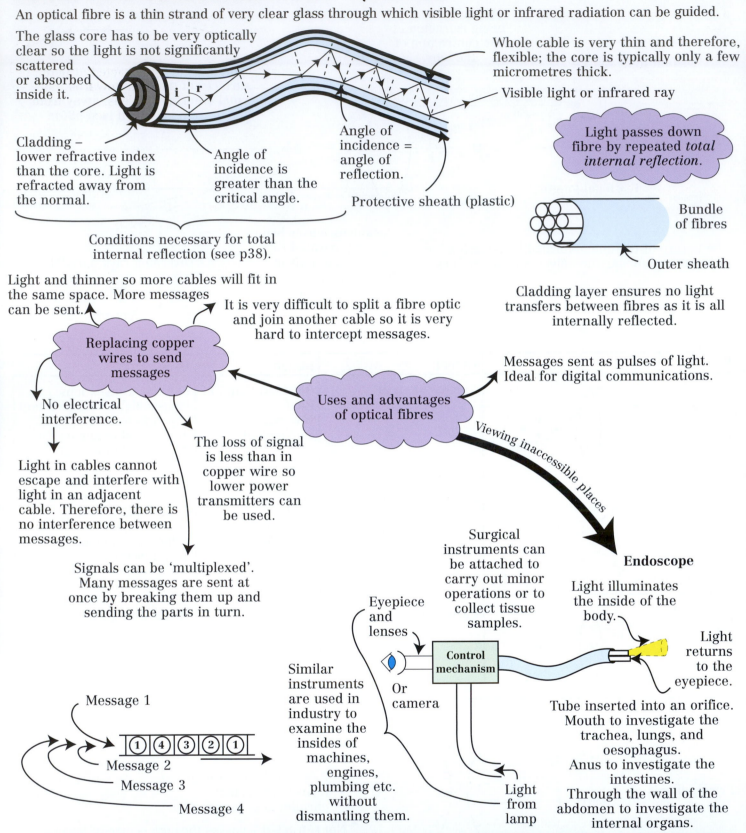

The glass core has to be very optically clear so the light is not significantly scattered or absorbed inside it.

Whole cable is very thin and therefore, flexible; the core is typically only a few micrometres thick.

Visible light or infrared ray

Angle of incidence = angle of reflection.

Cladding – lower refractive index than the core. Light is refracted away from the normal.

Angle of incidence is greater than the critical angle.

Conditions necessary for total internal reflection (see p38).

Protective sheath (plastic)

Light passes down fibre by repeated *total internal reflection*.

Bundle of fibres

Outer sheath

Cladding layer ensures no light transfers between fibres as it is all internally reflected.

Light and thinner so more cables will fit in the same space. More messages can be sent.

It is very difficult to split a fibre optic and join another cable so it is very hard to intercept messages.

Messages sent as pulses of light. Ideal for digital communications.

Replacing copper wires to send messages

Uses and advantages of optical fibres

Viewing inaccessible places

No electrical interference.

Light in cables cannot escape and interfere with light in an adjacent cable. Therefore, there is no interference between messages.

The loss of signal is less than in copper wire so lower power transmitters can be used.

Signals can be 'multiplexed'. Many messages are sent at once by breaking them up and sending the parts in turn.

Endoscope

Light illuminates the inside of the body.

Light returns to the eyepiece.

Surgical instruments can be attached to carry out minor operations or to collect tissue samples.

Eyepiece and lenses

Control mechanism

Or camera

Similar instruments are used in industry to examine the insides of machines, engines, plumbing etc. without dismantling them.

Light from lamp

Tube inserted into an orifice. Mouth to investigate the trachea, lungs, and oesophagus. Anus to investigate the intestines. Through the wall of the abdomen to investigate the internal organs.

Message 1
Message 2
Message 3
Message 4

| 1 | 4 | 3 | 2 | 1 |

Questions
1. Copy and complete the following diagram as accurately as possible showing the path of the light along the fibre-optic cable. What can you say about the size of the pairs of angles a and b, and x and y?
2. What types of electromagnetic radiation are commonly used with fibre optics?
3. Outline some benefits of using fibre optics rather than copper wires for sending messages.
4. The light in a fibre optic gradually gets less intense as it travels along the fibre due to impurities in the glass absorbing some of the light energy. What is the electrical equivalent of this?
5. What is an endoscope? Suggest two possible uses for one.
6. Suggest why doctors often prefer to see inside people using an endoscope rather than carrying out an operation to open up the patient.

WAVES AND COMMUNICATIONS Ultrasound and its Applications

Ultrasound is a sound wave with a frequency of greater than 20 000 Hz. This is above the upper limit of hearing for humans, so we cannot hear it, although in all other respects it behaves in exactly the same manner as normal sound.

Ultrasound can be used to detect the distance between the boundaries of two objects.

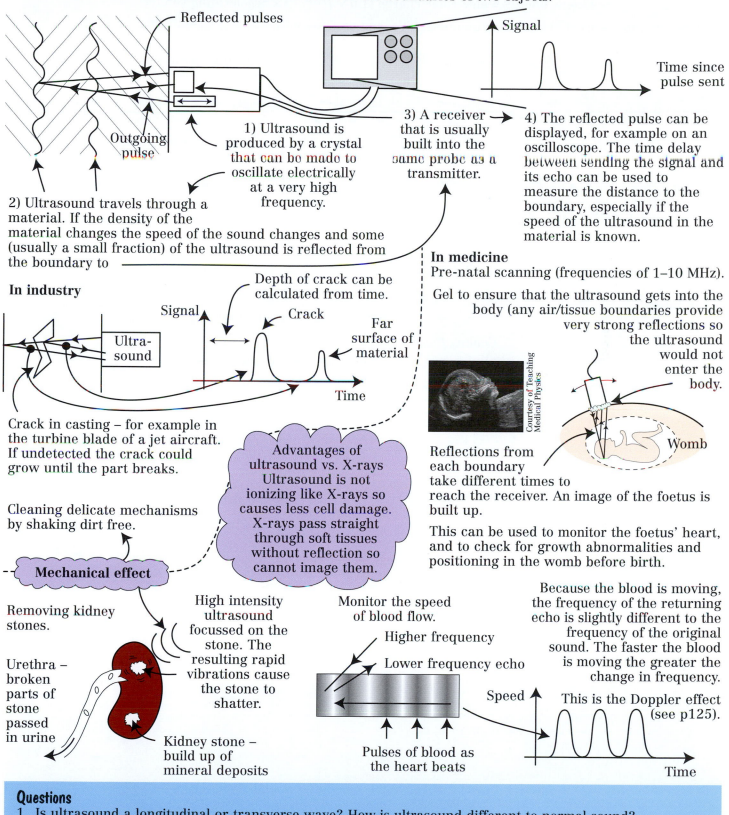

Reflected pulses

Outgoing pulse

Signal

Time since pulse sent

1) Ultrasound is produced by a crystal that can be made to oscillate electrically at a very high frequency.

2) Ultrasound travels through a material. If the density of the material changes the speed of the sound changes and some (usually a small fraction) of the ultrasound is reflected from the boundary to

3) A receiver that is usually built into the same probe as a transmitter.

4) The reflected pulse can be displayed, for example on an oscilloscope. The time delay between sending the signal and its echo can be used to measure the distance to the boundary, especially if the speed of the ultrasound in the material is known.

In industry

Signal

Depth of crack can be calculated from time.

Crack

Far surface of material

Ultra-sound

Time

Crack in casting – for example in the turbine blade of a jet aircraft. If undetected the crack could grow until the part breaks.

Cleaning delicate mechanisms by shaking dirt free.

Mechanical effect

Removing kidney stones.

Urethra – broken parts of stone passed in urine

Kidney stone – build up of mineral deposits

Advantages of ultrasound vs. X-rays
Ultrasound is not ionizing like X-rays so causes less cell damage. X-rays pass straight through soft tissues without reflection so cannot image them.

High intensity ultrasound focussed on the stone. The resulting rapid vibrations cause the stone to shatter.

In medicine
Pre-natal scanning (frequencies of 1–10 MHz).

Gel to ensure that the ultrasound gets into the body (any air/tissue boundaries provide very strong reflections so the ultrasound would not enter the body.

Courtesy of Teaching Medical Physics

Womb

Reflections from each boundary take different times to reach the receiver. An image of the foetus is built up.

This can be used to monitor the foetus' heart, and to check for growth abnormalities and positioning in the womb before birth.

Monitor the speed of blood flow.

Higher frequency

Lower frequency echo

Speed

Pulses of blood as the heart beats

Because the blood is moving, the frequency of the returning echo is slightly different to the frequency of the original sound. The faster the blood is moving the greater the change in frequency.

This is the Doppler effect (see p125).

Time

(see p125).

Questions
1. Is ultrasound a longitudinal or transverse wave? How is ultrasound different to normal sound?
2. The speed of ultrasound in soft tissue is 1540 m/s. The oscilloscope trace shows the returning pulses. How far below the surface of the body was pulse A and pulse B reflected?
3. Suggest two reasons why ultrasound may be preferable to X-rays for medical examinations.
4. Explain how ultrasound could be used to locate the depth below the skin of a cyst (fluid filled pocket) in an organ.
5. Suggest one use of ultrasound in medicine and one in industry other than for making images of hidden objects.

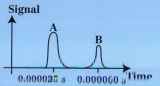

Signal

A

B

0.0000025 s   0.000060 s   Time

WAVES AND COMMUNICATIONS   Uses of Electron Beams

Review p57. Particularly note . . .
1. Electron beams are produced by 'boiling' electrons off a heated filament (thermionic emission). The hotter the filament the more electrons are produced.
2. The electrons are accelerated across a potential difference to increase their kinetic energy.
   Kinetic energy = electronic charge $(1.6 \times 10^{-19}$ C) × accelerating voltage

Cathode ray tubes – used in computer monitors, TVs, and oscilloscopes.

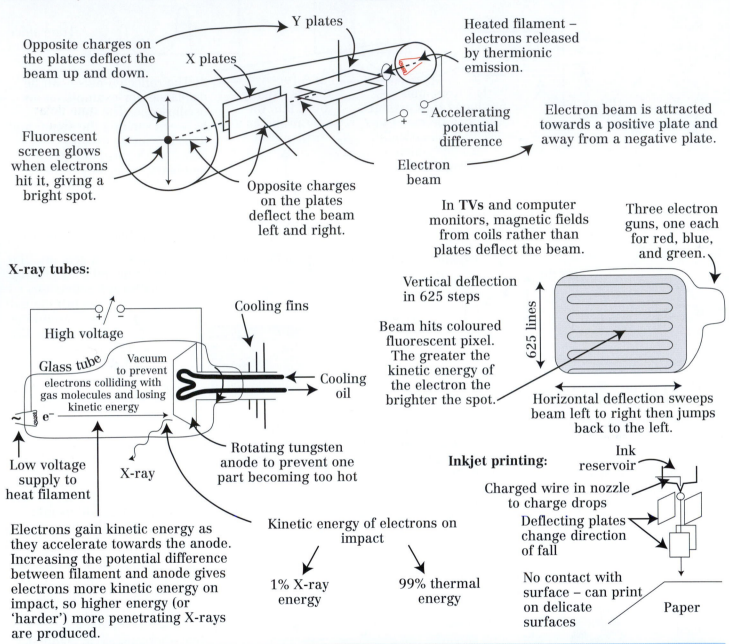

Opposite charges on the plates deflect the beam up and down.

Y plates

X plates

Heated filament – electrons released by thermionic emission.

Fluorescent screen glows when electrons hit it, giving a bright spot.

Accelerating potential difference

Electron beam is attracted towards a positive plate and away from a negative plate.

Opposite charges on the plates deflect the beam left and right.

Electron beam

In **TVs** and computer monitors, magnetic fields from coils rather than plates deflect the beam.

Three electron guns, one each for red, blue, and green.

Vertical deflection in 625 steps

625 lines

Beam hits coloured fluorescent pixel. The greater the kinetic energy of the electron the brighter the spot.

Horizontal deflection sweeps beam left to right then jumps back to the left.

X-ray tubes:

Cooling fins

High voltage

Glass tube

Vacuum to prevent electrons colliding with gas molecules and losing kinetic energy

Cooling oil

e^-

Low voltage supply to heat filament

X-ray

Rotating tungsten anode to prevent one part becoming too hot

Inkjet printing:

Ink reservoir

Charged wire in nozzle to charge drops

Deflecting plates change direction of fall

No contact with surface – can print on delicate surfaces

Paper

Electrons gain kinetic energy as they accelerate towards the anode. Increasing the potential difference between filament and anode gives electrons more kinetic energy on impact, so higher energy (or 'harder') more penetrating X-rays are produced.

Kinetic energy of electrons on impact

1% X-ray energy

99% thermal energy

Questions
1. The diagram shows the X and Y plates in an oscilloscope viewed end on. In each case which of the dots shown (a, b, or c) correctly shows the position of the beam falling on the screen?
2. How many lines are there on a TV screen? Explain how the electron beam is made to move across the screen.
3. Describe three ways that the tungsten anode in an X-ray tube is kept cool.
4. What adjustment to an X-ray tube produces X-rays that are more penetrating?
5. An X-ray tube accelerates an electron through a potential difference of 40 000 000 V. (Charge on the electron = 1.6×10^{-19}C.)
   a. Show that its kinetic energy when it hits the anode is about 6.4×10^{-12} J.
   b. If 1.6×10^{15} electrons hit the anode, show the total energy they deliver is about 10 kJ.
   c. If this energy is delivered in about 0.2 s what is the power of the tube?
   d. What percentage of the energy above is converted to X-ray energy and hence explain why the tungsten anode needs to be cooled?
   e. Explain what effect increasing the filament temperature would have on the number of X-rays produced in an X-ray tube.

WAVES AND COMMUNICATIONS Beams of Light – CDs and Relativity

Einstein's theory of relativity is one of the most creative and challenging ideas in physics, while reading the information from a CD is a very straightforward application of physics. Yet they both involve ideas about beams of light.

A beam of laser light reads the information stored on a CD (or DVD).

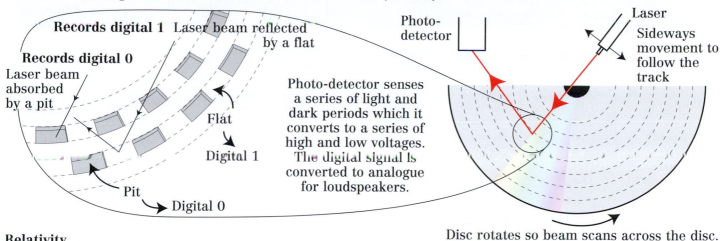

Records digital 1 Laser beam reflected by a flat

Records digital 0
Laser beam absorbed by a pit

Flat → Digital 1

Pit → Digital 0

Photo-detector

Laser
Sideways movement to follow the track

Photo-detector senses a series of light and dark periods which it converts to a series of high and low voltages. The digital signal is converted to analogue for loudspeakers.

Disc rotates so beam scans across the disc.

Relativity

This theory makes some weird predictions about how we measure length and time when moving very fast relative to another object.

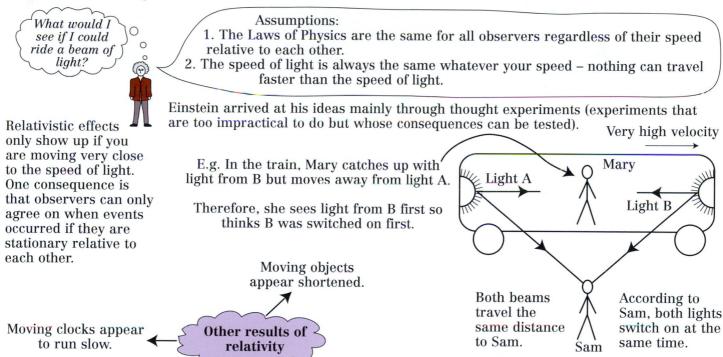

What would I see if I could ride a beam of light?

Assumptions:
1. The Laws of Physics are the same for all observers regardless of their speed relative to each other.
2. The speed of light is always the same whatever your speed – nothing can travel faster than the speed of light.

Relativistic effects only show up if you are moving very close to the speed of light. One consequence is that observers can only agree on when events occurred if they are stationary relative to each other.

Einstein arrived at his ideas mainly through thought experiments (experiments that are too impractical to do but whose consequences can be tested).

Very high velocity

E.g. In the train, Mary catches up with light from B but moves away from light A.

Therefore, she sees light from B first so thinks B was switched on first.

Light A   Mary   Light B

Both beams travel the same distance to Sam.

According to Sam, both lights switch on at the same time.

Sam

Moving objects appear shortened.

Moving clocks appear to run slow.

Other results of relativity

The mass of an object appears to increase the faster it travels. This leads to the famous equation $\Delta E = \Delta mc^2$.

Some scientists did not like Einstein's ideas because they suggested Newton's Laws were not exactly right.

Other scientists

Atomic clocks flown around the world on a jet aircraft record slightly less time than a stationary clock left on Earth.

Cosmic rays create short-lived particles in the atmosphere. These travel towards the Earth's surface and should decay before reaching it, but do not as the distance appears much shorter to them. Therefore, they can easily cover it in their lifetime.

Tested the predictions of relativity

The extraction of energy by nuclear fission and fusion relies on $\Delta E = \Delta mc^2$ being correct.

Questions
1. Laser beams can be made very narrow and do not spread out much. Why is this necessary for reading a CD as described above?
2. If you shake a CD player while playing a disc the music can be interrupted or skip a section. Using the above description try to explain why.
3. What is a thought experiment?
4. What predictions did Einstein make from his thought experiments?
5. Suggest three ways Einstein's predictions have been tested.

RADIOACTIVITY How is Nuclear Radiation Used in Hospitals?

Stable nuclei are bombarded with protons. These unstable proton-rich nuclei decay by beta-plus emission with short half-lives. They emit positrons.

Positron emitter made into a drug (designed to collect quickly in the organ of interest) and injected into the patient.

Especially if it is a cancerous tumour

Positron emission tomography (PET)

4. Gamma rays travel off in opposite directions to conserve momentum.

5. Gamma ray pairs are detected by circular detectors, which give a good indication of their origin.

1. Unstable nucleus undergoes beta-plus decay and emits a positron (e⁺).

2. Positron travels about 1 mm before meeting an electron.

1 mm

3. The two particles annihilate each other and become two gamma rays.

6. The origin of the gamma rays shows where the positron emitting drug has collected.

This can be used to find out how well the drug moves round the body and how well organs of interest are working, or if they contain a tumour.

Progress of the drug can be tracked by a radiation detector outside the body.

Radioisotope attached to a drug that is absorbed by an organ of interest, e.g. a kidney.

Monitoring the flow of the radioisotope over time can tell doctors about how organs are working.

Blockage – radioisotope will not pass and no radiation detected in this area outside body.

Gamma emitter made into a drug and injected.

Short half-life used so radiation does not stay in the body too long.

Alpha emitters cannot be used as they would not pass through the body and are highly ionizing so would cause a lot of cell damage.

Radioisotope Tracers

PET Scanner

HOSPITAL

X-ray Department

Cancer Ward

Cancerous cells are those where the DNA has been damaged and grow and divide uncontrollably.

Ionizing radiation can damage the DNA in cells.

If the radiation dose is small, the cells may be able to repair themselves. Large doses of radiation can kill cells. This can be used to kill cancerous cells.

X-rays are high frequency, short wavelength, electromagnetic waves. They are ionizing so can damage cells. Exposure to them needs to be limited.

The benefits of the use of X-rays to diagnose medical problems often outweigh any cell damage caused.

X-rays expose photographic film and bones show up as a shadow.

Intensity of X-ray decreases.

Bone contains more heavy atoms, e.g. calcium, which absorbs X-rays strongly.

Flesh contains lighter atoms that do not absorb X-rays strongly.

Source of gamma rays

Source rotated around patient centred on the tumour.

Cells around the tumour receive less radiation – they should recover.

Cancerous tumour – radiation most concentrated here. These cells should be killed.

Radiotherapy may not be successful if the tumour is very large. In this case radiotherapy may be used to reduce suffering. This is called *palliative* care.

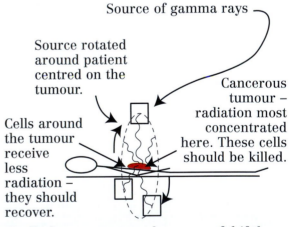

Courtesy of Teaching Medical Physics

Questions
1. Which types of radiation, alpha, beta, or gamma can pass through flesh?
2. Why do the radioisotopes injected into patients always have short half-lives?
3. What absorbs X-rays better, flesh or bone?
4. What does PET stand for? Describe how it works, for example to identify the location of a cancerous tumour.
5. The thyroid gland stores iodine. How could injecting a patient with radioactive iodine-123 allow a doctor to find out how well the thyroid gland is working?
6. Ionizing radiation can cause the DNA in cells to mutate and cause cancer. Therefore, why can we also use ionizing radiation as a treatment for cancer?
7. Why is the source of gamma rays in radiotherapy rotated around the patient?
8. All ionizing radiation causes damage to the body. How do doctors justify exposing patients to it?

RADIOACTIVITY Other Uses of Radioactivity

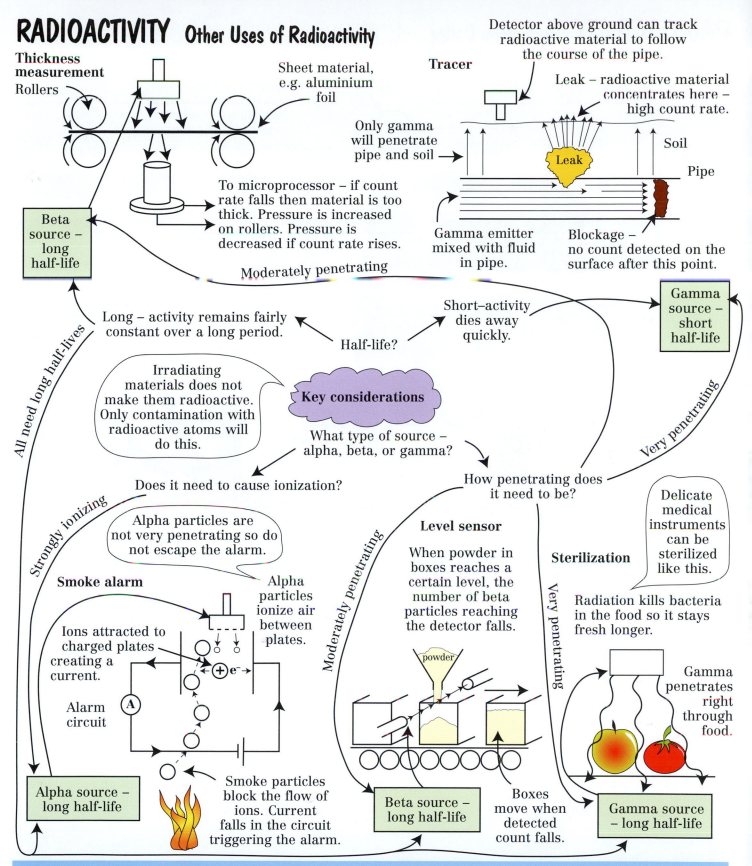

Questions
1. Explain whether an alpha, beta, or gamma source is most useful for the following and why:
   a. Smoke alarms.
   b. Detecting aluminium foil thickness in a factory.
   c. Following the flow of oil along a pipe.
2. Should a radioactive material with a long or short half-life be chosen for the following and why?
   a. Smoke alarm.
   b. Tracer in an oil pipe.
   c. Thickness detection in a factory.
3. Many people are concerned about the effect on their health of radioactive sources. How would you address the following concerns?
   a. 'I don't have a smoke alarm as I do not want a radioactive source in my house.'
   b. 'I am concerned that irradiated food might be radioactive.'

RADIOACTIVITY   Radioactive Dating

Carbon dating

Cosmic rays from space hit carbon atoms in carbon dioxide in the atmosphere and convert a very small number of them (about four in every three million, million) to carbon-14.

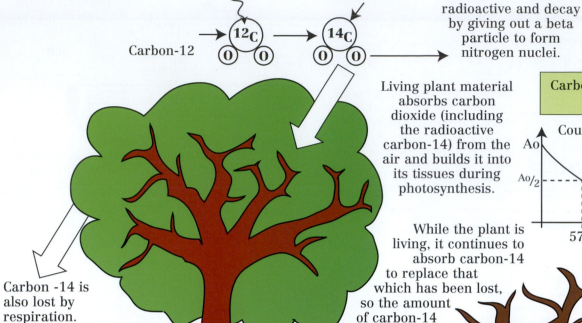

Cosmic ray

Carbon-12 ^{12}C O   O   Carbon-14 ^{14}C O   O

Carbon-14 nuclei are radioactive and decay by giving out a beta particle to form nitrogen nuclei.

$$^{14}_{6}C \rightarrow \, ^{14}_{7}N = \, ^{0}_{-1}\beta$$

Living plant material absorbs carbon dioxide (including the radioactive carbon-14) from the air and builds it into its tissues during photosynthesis.

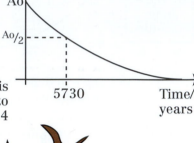

Carbon-14 has a half-life of 5730 years.

Counts

A_0

$A_0/2$

5730   Time/years

Carbon -14 is also lost by respiration.

While the plant is living, it continues to absorb carbon-14 to replace that which has been lost, so the amount of carbon-14 in its tissues remains constant.

When the plant dies, no more carbon-14 is absorbed and the carbon-14 in the plant's tissues begins to decay away.

Measuring the activity of a sample of ancient materials that were once living and comparing the activity to a living sample can give a fairly accurate indication of when the ancient material was alive.

(It works for plant or animal material because animals eat plants and absorb carbon-14 from them.)

Assumption: the concentration of $^{14}CO_2$ in the atmosphere has remained constant.

Very small quantities are involved leading to significant uncertainties.

Dating of rocks
Many rocks contain traces of radioactive uranium. This decays to stable lead with a half-life of 4.5 billion years.

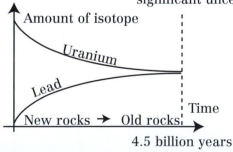

Amount of isotope

Uranium

Lead

New rocks → Old rocks   Time

4.5 billion years

Assuming that there was no lead in the rock when it was formed the ratio of uranium to lead gives an approximate age for the rock.

Questions
1. The graph shows the radioactive decay of carbon-14.
   a. Use the graph to calculate the half-life of carbon-14. What does carbon-14 decay into?
   b. A wooden post from an archaeological dig produces 150 counts/min. Wood from an identical species of tree currently alive gives 600 counts/min. How long ago did the wood from the archaeological dig die?
   c. What assumption have you made in the above calculation?
2. Two samples of rock are analysed. The ratio of 238-uranium to 206-lead are as follows:
   Sample A: uranium to lead 5:1   Sample B: uranium to lead 7:1.
   Which rock is older and how do you know? What assumption have you made?
3. The age of the Earth is thought to be about 4.5 billion years. Why can there be no rock in which the number of lead nuclei formed from the decay of uranium outweighs the number of uranium nuclei remaining?

Decay of carbon-14

No. of counts: 300, 250, 200, 150, 100, 50, 0

Time/years: 2000, 4000, 6000, 8000, 10 000, 12 000, 14 000, 16 000, 18 000, 20 000

RADIOACTIVITY Nuclear Power and Weapons

See p77 for a description of the nuclear fission process and the nuclear reactor.

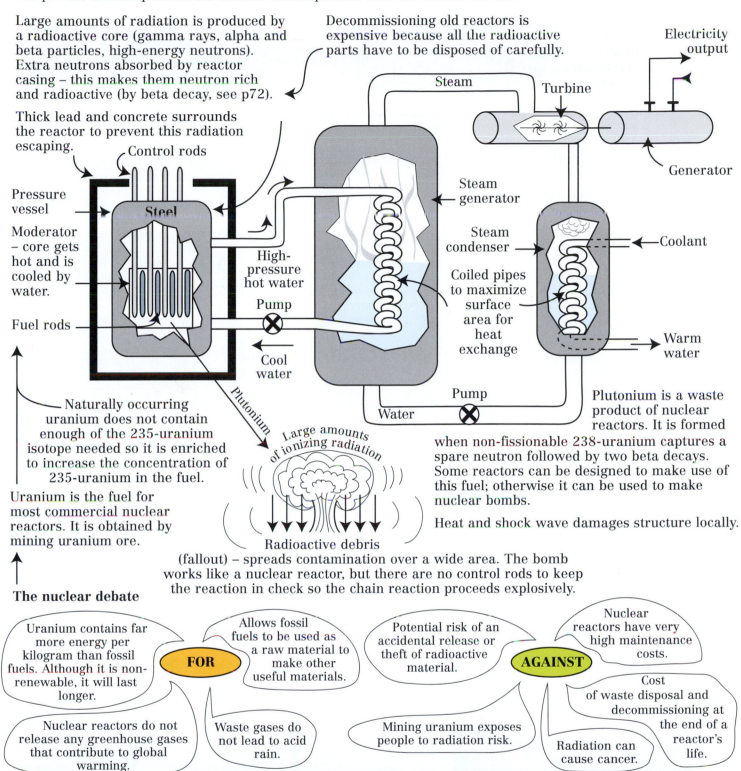

Large amounts of radiation is produced by a radioactive core (gamma rays, alpha and beta particles, high-energy neutrons). Extra neutrons absorbed by reactor casing – this makes them neutron rich and radioactive (by beta decay, see p72).

Thick lead and concrete surrounds the reactor to prevent this radiation escaping.

Decommissioning old reactors is expensive because all the radioactive parts have to be disposed of carefully.

Control rods

Pressure vessel

Moderator – core gets hot and is cooled by water.

Fuel rods

Steel

High-pressure hot water

Pump

Cool water

Plutonium

Steam

Turbine

Electricity output

Generator

Steam generator

Steam condenser

Coiled pipes to maximize surface area for heat exchange

Coolant

Warm water

Pump

Water

Naturally occurring uranium does not contain enough of the 235-uranium isotope needed so it is enriched to increase the concentration of 235-uranium in the fuel.

Uranium is the fuel for most commercial nuclear reactors. It is obtained by mining uranium ore.

Large amounts of ionizing radiation

Radioactive debris (fallout) – spreads contamination over a wide area. The bomb works like a nuclear reactor, but there are no control rods to keep the reaction in check so the chain reaction proceeds explosively.

Plutonium is a waste product of nuclear reactors. It is formed when non-fissionable 238-uranium captures a spare neutron followed by two beta decays. Some reactors can be designed to make use of this fuel; otherwise it can be used to make nuclear bombs.

Heat and shock wave damages structure locally.

The nuclear debate

Uranium contains far more energy per kilogram than fossil fuels. Although it is non-renewable, it will last longer.

Nuclear reactors do not release any greenhouse gases that contribute to global warming.

FOR

Allows fossil fuels to be used as a raw material to make other useful materials.

Waste gases do not lead to acid rain.

Potential risk of an accidental release or theft of radioactive material.

AGAINST

Nuclear reactors have very high maintenance costs.

Cost of waste disposal and decommissioning at the end of a reactor's life.

Mining uranium exposes people to radiation risk.

Radiation can cause cancer.

Questions

1. Write out a list of energy changes in a nuclear power station starting from nuclear energy stored in uranium fuel and ending with electrical energy in the wires leading from the generator.
2. Where does the fuel for a nuclear power station come from and what has to happen to it before it can be used?
3. The energy released by 1 kg of ^{235}U is about 8×10^{13} J. Show that this could light a 60 W light bulb for about 42 thousand years.
4. Using the diagram of a nuclear power plant above explain:
   a. Why is the reactor surrounded by a thick layer of concrete and lead?
   b. Why is the pressure vessel made of steel?
   c. Why are the pipes in the heat exchangers coiled up?
5. Nuclear weapons cause damage to living things in three ways – what are they?
6. 'Nuclear power damages the environment and should be banned.' Give arguments in favour and against this statement.

RADIOACTIVITY Radioactive Waste

Sources:
- Nuclear fission power stations (p111).
- Industrial users of radioactivity (p109).
- Hospitals and other medical establishments (p108).
- Laboratories.
- Decommissioned nuclear weapons.

These wastes should be disposed of in a way that does not significantly increase the naturally occurring background level of radiation around the disposal site.

Waste is classified into three levels by considering:

- How long the waste will remain at a hazardous level.
- The concentration of radioactive material in the waste.
- Whether it is heat generating.

Low-level

Waste paper, rags, clothing, filters.

Mainly very small amounts of short half-life isotopes.

Enough radioactive material to require action to protect people but not enough to require special handling or storage. The definition of a 'safe' level might change over time.

Intermediate level

Materials that have been in direct contact with highly radioactive isotopes, e.g. nuclear fuel cladding.

Short half-life   Long half-life

High-level

Radioactive isotopes, e.g. fission products from a reactor. 95% of total radioactivity but a very small volume.

Requires both shielding and cooling.

It is highly radioactive and hot as nuclear decay is still occurring at a high rate.

Allowed to cool under water for about 3 months.

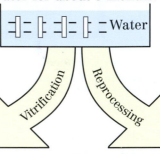 Water

Vitrification   Reprocessing

Half-life >30 years or a high proportion of alpha emitters

Protects against gamma rays

Comprises about 90% of the volume but 1% of the radioactivity of all radioactive waste.

Requires shielding by encapsulating in concrete (and sometimes lead).

Waste is mixed with glass (which is chemically unreactive and insoluble). Helps to prevent waste leaking out.

Some spent fuel still contains unreacted isotopes that are in too low a concentration to be useful. They are extracted, concentrated, and added to new reactor fuel.

Concrete

Waste

Steel or copper drum

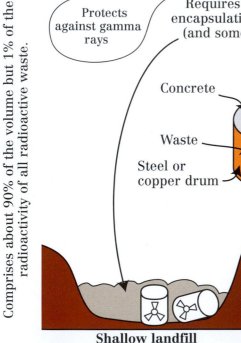

Shallow landfill

Can be difficult to site as the local population may have concerns about their safety.

Eventually the store will be filled with concrete and sealed when the waste has cooled enough and the store is full.

Stable rocks – few cracks that to run through the store and radioactive material into the

would allow water potentially carry groundwater.

Store must be secure to prevent radioactive materials falling into the wrong hands, e.g. terrorist organizations.

Air circulated by fans to remove heat produced by the still decaying waste.

Managed underground store

Questions
1. What are the three classifications of nuclear waste?
2. What types of materials make up low-level waste?
3. What is the main constituent of intermediate level waste?
4. What constitutes high-level waste and why is this generally hot?
5. What happens to low-level waste?
6. What happens to intermediate level waste?
7. What happens to high-level waste?
8. Why are spent fuel rods left in cooling ponds for 3 months after use?
9. You are responsible for finding a site for a new managed underground radioactive waste store.
   a. What features would you look for in identifying a suitable site?
   b. What concerns might local residents have?
   c. How might you go about addressing these concerns?

OUR PLACE IN THE UNIVERSE Geological Processes

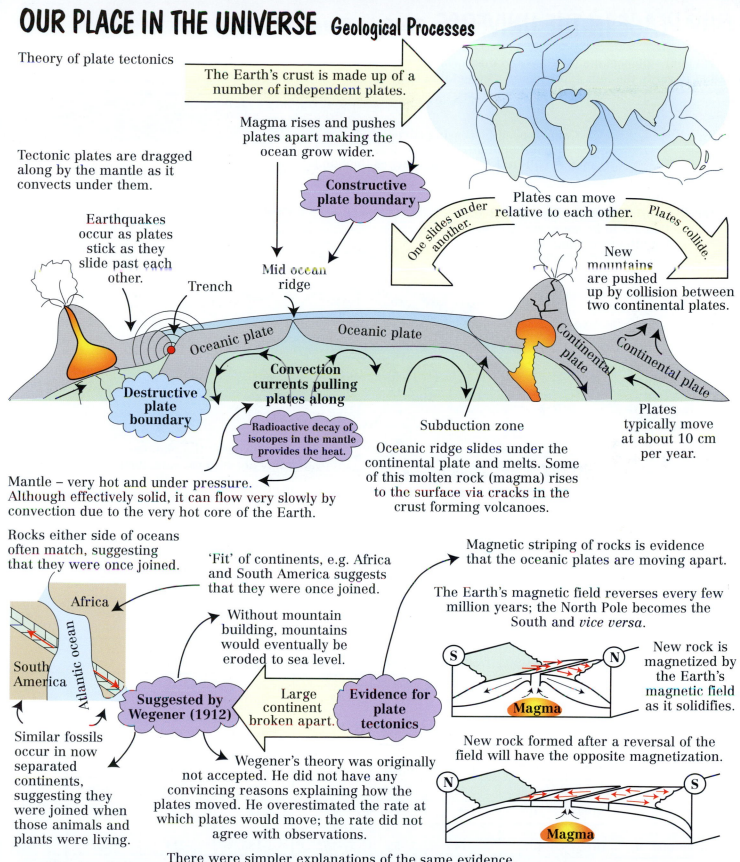

Theory of plate tectonics

The Earth's crust is made up of a number of independent plates.

Magma rises and pushes plates apart making the ocean grow wider.

Constructive plate boundary

Tectonic plates are dragged along by the mantle as it convects under them.

Plates can move relative to each other.

One slides under another.

Plates collide.

New mountains are pushed up by collision between two continental plates.

Earthquakes occur as plates stick as they slide past each other.

Trench

Mid ocean ridge

Oceanic plate

Oceanic plate

Continental plate

Continental plate

Destructive plate boundary

Convection currents pulling plates along

Radioactive decay of isotopes in the mantle provides the heat.

Subduction zone

Plates typically move at about 10 cm per year.

Oceanic ridge slides under the continental plate and melts. Some of this molten rock (magma) rises to the surface via cracks in the crust forming volcanoes.

Mantle – very hot and under pressure. Although effectively solid, it can flow very slowly by convection due to the very hot core of the Earth.

Rocks either side of oceans often match, suggesting that they were once joined.

'Fit' of continents, e.g. Africa and South America suggests that they were once joined.

Magnetic striping of rocks is evidence that the oceanic plates are moving apart.

The Earth's magnetic field reverses every few million years; the North Pole becomes the South and *vice versa*.

Africa

Atlantic ocean

South America

Without mountain building, mountains would eventually be eroded to sea level.

Suggested by Wegener (1912)

Large continent broken apart.

Evidence for plate tectonics

New rock is magnetized by the Earth's magnetic field as it solidifies.

S **N** **Magma**

Similar fossils occur in now separated continents, suggesting they were joined when those animals and plants were living.

Wegener's theory was originally not accepted. He did not have any convincing reasons explaining how the plates moved. He overestimated the rate at which plates would move; the rate did not agree with observations.

New rock formed after a reversal of the field will have the opposite magnetization.

N **S** **Magma**

There were simpler explanations of the same evidence.

Questions
1. What is the difference between a constructive and destructive plate boundary?
2. Explain why the majority of earthquakes and volcanoes occur near plate boundaries.
3. Give three pieces of evidence mentioned above in support of the idea of plate tectonics.
4. Why did people find it difficult to accept Wegener's ideas?
5. What is the name of the process that causes the material in the mantle to circulate and drag the plates along?
6. Describe how the magnetization of the rocks of the oceanic crust could be used to show that the ocean is growing wider over millions of years.
7. Describe and explain the differences between the collision of two continental plates compared to a continental and an oceanic plate.

OUR PLACE IN THE UNIVERSE The Solar System

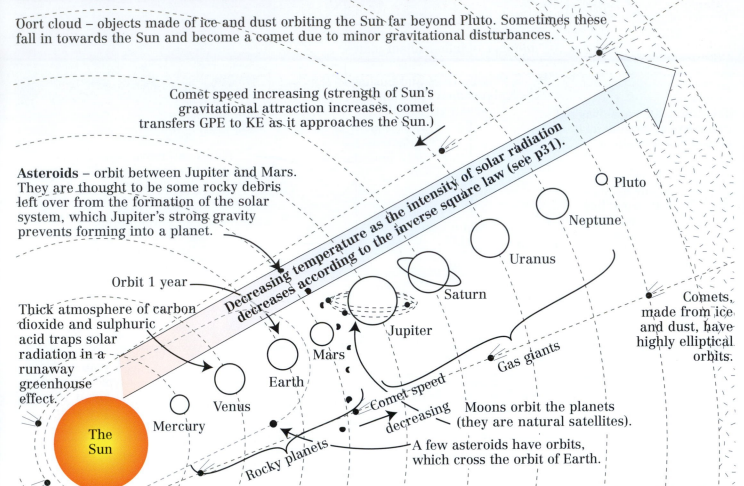

Oort cloud – objects made of ice and dust orbiting the Sun far beyond Pluto. Sometimes these fall in towards the Sun and become a comet due to minor gravitational disturbances.

Comet speed increasing (strength of Sun's gravitational attraction increases, comet transfers GPE to KE as it approaches the Sun.)

Asteroids – orbit between Jupiter and Mars. They are thought to be some rocky debris left over from the formation of the solar system, which Jupiter's strong gravity prevents forming into a planet.

Decreasing temperature as the intensity of solar radiation decreases according to the inverse square law (see p31).

Orbit 1 year

Thick atmosphere of carbon dioxide and sulphuric acid traps solar radiation in a runaway greenhouse effect.

The Sun

Mercury

Venus

Earth

Mars

Jupiter

Saturn

Uranus

Neptune

Pluto

Gas giants

Comet speed decreasing

Rocky planets

Comets, made from ice and dust, have highly elliptical orbits.

Moons orbit the planets (they are natural satellites).

A few asteroids have orbits, which cross the orbit of Earth.

The planets orbit the Sun in elliptical (slightly squashed circle) orbits. The Sun is at one focus of the ellipse.

Quantity	Mercury	Venus	Earth	Mars	Jupiter	Saturn	Uranus	Neptune	Pluto
Mean distance from sun (orbit radius), million km	57.9	108	150	228	778	1430	2870	4500	5900
Time to orbit the Sun, years	0.241	0.615	1.00	1.88	11.9	29.5	84.0	165	248
Orbital speed, km/s	47.9	35.0	29.8	24.1	13.1	9.64	6.81	5.43	4.74
Equatorial diameter, km	4880	12 100	12 800	6790	143 000	120 000	51 800	49 500	?3000
Mass (Earth = 1)	0.0558	0.815	1.000	0.107	318	95.1	14.5	17.2	?0.010
Density g/cm^3	5.600	5.200	5.520	3.950	1.310	0.704	1.210	1.670	?
Moons	0	0	1	2	16	17	15	2	1
Typical surface temperature, ºC	167	457	14	−55	−153	−185	−214	−225	−236
Atmosphere	None	Carbon dioxide	Nitrogen, oxygen	Carbon dioxide	Hydrogen, helium	Hydrogen, helium	Hydrogen, helium	Hydrogen, helium	None

Questions
1. Which of the following orbit the Sun directly and which orbit planets?
   Comets, moons, asteroids, artificial satellites, planets.
2. Explain why the density of Jupiter, Saturn, Uranus, and Neptune is a lot less than that of Mercury, Venus, Earth, and Mars.
3. Using the data in the table show that: a. The circumference of the Earth's orbit is 942 million km.
   b. The time the Earth takes to orbit the Sun is 31.6 × 10^6 s.   c. That 31.6 × 10^6 s = 1 year.
4. Plot a graph of surface temperature vs. distance from the Sun. State and explain any trend you see. One planet is anomalous, which is it and give a scientific explanation for why it does not fit the trend?
5. Explain why the speed of a comet decreases as it moves away from the Sun.

OUR PLACE IN THE UNIVERSE
Telescopes and Types of Radiation Used to Learn About the Universe

Everything we know about space outside the solar system comes from analyzing the electromagnetic radiation collected from space by telescopes.
Different objects in space emit different wavelengths.

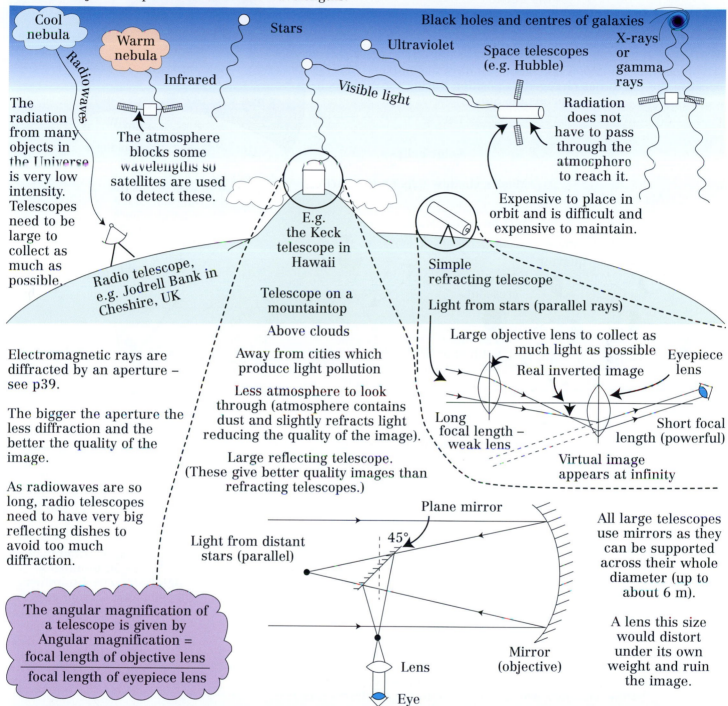

Cool nebula

Warm nebula

Radiowaves

Infrared

Stars

Visible light

Ultraviolet

Black holes and centres of galaxies

Space telescopes (e.g. Hubble)

X-rays or gamma rays

The radiation from many objects in the Universe is very low intensity. Telescopes need to be large to collect as much as possible.

The atmosphere blocks some wavelengths so satellites are used to detect these.

Radiation does not have to pass through the atmosphere to reach it.

Expensive to place in orbit and is difficult and expensive to maintain.

Radio telescope, e.g. Jodrell Bank in Cheshire, UK

E.g. the Keck telescope in Hawaii

Telescope on a mountaintop

Above clouds

Simple refracting telescope

Light from stars (parallel rays)

Electromagnetic rays are diffracted by an aperture – see p39.

The bigger the aperture the less diffraction and the better the quality of the image.

As radiowaves are so long, radio telescopes need to have very big reflecting dishes to avoid too much diffraction.

Away from cities which produce light pollution

Less atmosphere to look through (atmosphere contains dust and slightly refracts light reducing the quality of the image).

Large reflecting telescope. (These give better quality images than refracting telescopes.)

Large objective lens to collect as much light as possible

Real inverted image

Eyepiece lens

Long focal length – weak lens

Short focal length (powerful)

Virtual image appears at infinity

Plane mirror

45°

Light from distant stars (parallel)

Lens

Eye

Mirror (objective)

All large telescopes use mirrors as they can be supported across their whole diameter (up to about 6 m).

A lens this size would distort under its own weight and ruin the image.

The angular magnification of a telescope is given by
$$\text{Angular magnification} = \frac{\text{focal length of objective lens}}{\text{focal length of eyepiece lens}}$$

see p39.

Questions
1. Make a list of the advantages and disadvantages of space telescopes compared to ground based telescopes.
2. Why do you think optical telescopes that collect visible light are often placed on mountains whilst radio telescopes can be at sea level?
3. Will the image in a refracting telescope be upright or inverted? Use a ray diagram to illustrate your answer. Suggest two advantages of having a very large objective lens and explain why there is a limit on how big the objective lens can be.
4. The aperture of a reflecting telescope is 0.7 m in diameter and it collects light of wavelength of about 0.00000055 m. Its objective mirror has a focal length of 0.4 m and its eyepiece a focal length of 1.50 cm. The diameter of the Jodrell Bank radio telescope dish is 76.2 m and the wavelengths it collects are around 1 m.
   a. What is the angular magnification of the reflecting telescope?
   b. Which telescope would you expect to suffer the most from diffraction?
5. Suggest at least three reasons why astronomers need to work together in international groups.

OUR PLACE IN THE UNIVERSE The Motion of Objects in the Sky

The Moon orbits the Earth once every 27.3 days. It also takes 27.3 days to rotate once so it always presents the same side to the Earth.

The Earth spins on its axis, through 360° once every 23 hrs 56 minutes.

This is a *sidereal day.*

The phases of the Moon take about 29 days to complete a cycle because the Earth has changed its position during the cycle as it orbits the Sun.

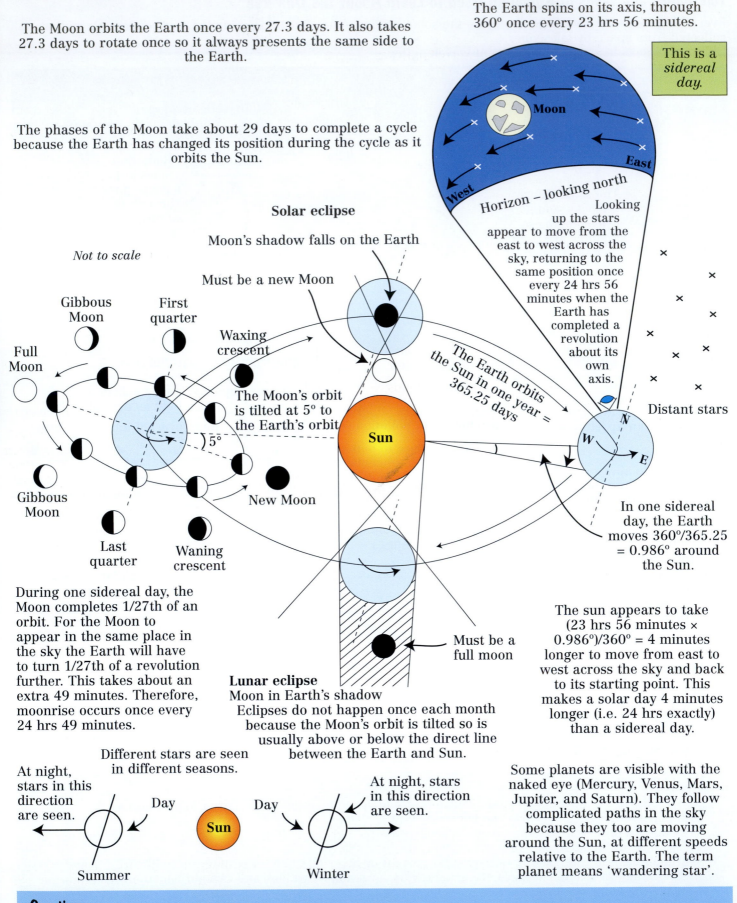

Solar eclipse

Moon's shadow falls on the Earth

Must be a new Moon

Not to scale

Gibbous Moon

First quarter

Full Moon

Waxing crescent

The Moon's orbit is tilted at 5° to the Earth's orbit

5°

Gibbous Moon

New Moon

Last quarter

Waning crescent

The Earth orbits the Sun in one year = 365.25 days

Sun

Must be a full moon

Lunar eclipse
Moon in Earth's shadow

Horizon – looking north

Looking up the stars appear to move from the east to west across the sky, returning to the same position once every 24 hrs 56 minutes when the Earth has completed a revolution about its own axis.

Moon

West

East

Distant stars

N

W

E

In one sidereal day, the Earth moves 360°/365.25 = 0.986° around the Sun.

During one sidereal day, the Moon completes 1/27th of an orbit. For the Moon to appear in the same place in the sky the Earth will have to turn 1/27th of a revolution further. This takes about an extra 49 minutes. Therefore, moonrise occurs once every 24 hrs 49 minutes.

Eclipses do not happen once each month because the Moon's orbit is tilted so is usually above or below the direct line between the Earth and Sun.

The sun appears to take (23 hrs 56 minutes × 0.986°)/360° = 4 minutes longer to move from east to west across the sky and back to its starting point. This makes a solar day 4 minutes longer (i.e. 24 hrs exactly) than a sidereal day.

Different stars are seen in different seasons.

At night, stars in this direction are seen.

Day

Sun

Day

At night, stars in this direction are seen.

Summer

Winter

Some planets are visible with the naked eye (Mercury, Venus, Mars, Jupiter, and Saturn). They follow complicated paths in the sky because they too are moving around the Sun, at different speeds relative to the Earth. The term planet means 'wandering star'.

Questions
1. Do most objects in the sky appear to move east to west or west to east? Which objects do not always follow this pattern?
2. What is the difference between a solar day and a sidereal day?
3. What is the difference between a lunar eclipse and a solar eclipse?
4. Why do we not have an eclipse once a month?
5. Why does the Moon appear at slightly different places in the sky each night at the same time?

OUR PLACE IN THE UNIVERSE Exploring Space

The Universe is vast. The light from the nearest star takes 4.2 years to reach Earth. If we could build a spaceship to travel at the speed of light the round trip journey time would be as follows.

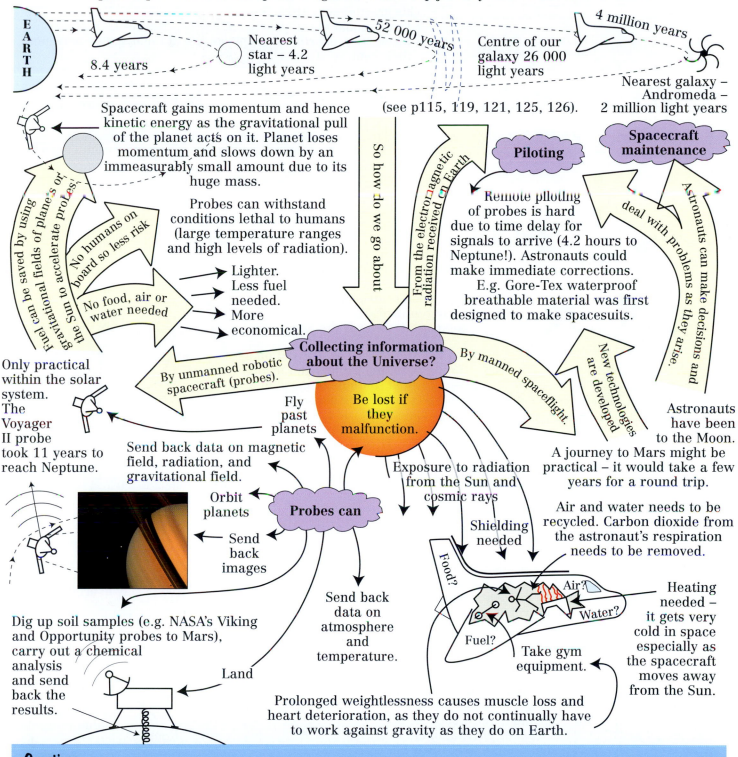

E A R T H

8.4 years

Nearest star – 4.2 light years

52 000 years

Centre of our galaxy 26 000 light years

4 million years

Nearest galaxy – Andromeda – 2 million light years

(see p115, 119, 121, 125, 126).

Spacecraft gains momentum and hence kinetic energy as the gravitational pull of the planet acts on it. Planet loses momentum and slows down by an immeasurably small amount due to its huge mass.

Probes can withstand conditions lethal to humans (large temperature ranges and high levels of radiation).

Fuel can be saved by using gravitational fields of planets or the Sun to accelerate probes.

No humans on board so less risk

No food, air or water needed

→ Lighter.
→ Less fuel needed.
→ More economical.

So how do we go about

From the electromagnetic radiation received on Earth

Piloting

Spacecraft maintenance

Remote piloting of probes is hard due to time delay for signals to arrive (4.2 hours to Neptune!). Astronauts could make immediate corrections. E.g. Gore-Tex waterproof breathable material was first designed to make spacesuits.

deal with problems as they arise

Astronauts can make decisions and

New technologies are developed

By manned spaceflight.

Collecting information about the Universe?

By unmanned robotic spacecraft (probes).

Only practical within the solar system. The Voyager II probe took 11 years to reach Neptune.

Be lost if they malfunction.

Fly past planets

Send back data on magnetic field, radiation, and gravitational field.

Orbit planets

Probes can

Send back images

Send back data on atmosphere and temperature.

Exposure to radiation from the Sun and cosmic rays

Shielding needed

Astronauts have been to the Moon.

A journey to Mars might be practical – it would take a few years for a round trip.

Air and water needs to be recycled. Carbon dioxide from the astronaut's respiration needs to be removed.

Food?

Air?

Water?

Fuel?

Take gym equipment.

Heating needed – it gets very cold in space especially as the spacecraft moves away from the Sun.

Dig up soil samples (e.g. NASA's Viking and Opportunity probes to Mars), carry out a chemical analysis and send back the results.

Land

Prolonged weightlessness causes muscle loss and heart deterioration, as they do not continually have to work against gravity as they do on Earth.

Questions
1. What are the three main ways scientists find out about the Universe?
2. Copy and complete the following table to summarize the advantages and disadvantages of two ways of exploring the solar system.

	Manned spaceflight	Unmanned robotic probes
Advantages		
Disadvantages		

3. It is proposed to send astronauts to Mars. Apart from the journey time of a couple of years, what other considerations are necessary when designing a spacecraft to make the journey?
4. 'Exploring Space is a waste of money that would be better spent on giving aid to people who live in poverty.' Do you agree or disagree with this statement? Give some explanation to try to convince somebody to support your view.
5. Explain why manned missions outside the solar system are very unlikely.

OUR PLACE IN THE UNIVERSE Forces in the Solar System

All masses exert gravitational attractions on all other masses.

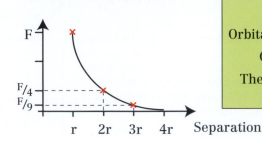

Equal

Pull of   Pull of
M on m   m on M

This gravitational attraction is:
• Proportional to the product of the two masses ($F \propto M \times m$)
• Inversely proportional to the square of the distance between their centres of mass ($F \propto 1/r^2$).

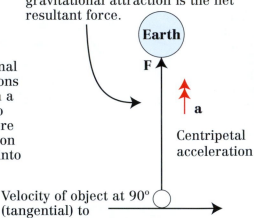

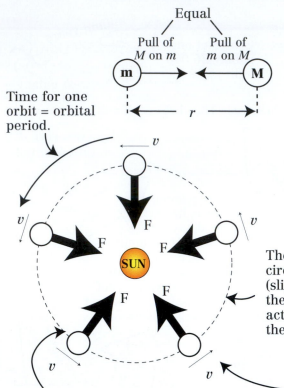

Time for one orbit = orbital period.

This centripetal force is provided by the gravitational attraction between the planet and the Sun.

Satellites and moons orbiting a planet also need centripetal force acting towards the planet at the centre of their orbit. It is provided by the gravitational attraction between the satellite or moon and the planet.

From p19 Centripetal force = mass × velocity²/radius of orbit.

Therefore, to stay in orbit at a particular distance from a larger body, a smaller body must travel at a particular speed in order that the centripetal force required is exactly provided by the available gravitational attraction between the masses.

Mathematics

Orbital circumference = $2\pi r$

Orbital period = T

Therefore orbital speed,
$v = 2\pi r/T$

The planets follow nearly circular paths around the Sun (slightly elliptical). To do so they need centripetal force acting towards the centre of their orbit, towards the Sun.

If there were no gravitational forces, the planets and moons would continue to move in a straight line according to Newton's First Law, as there would be no forces acting on them. They would drift off into space.

Orbiting body

Centripetal force provided by gravitational attraction is the net resultant force.

Earth

a

Centripetal acceleration

Velocity of object at 90° (tangential) to acceleration.

As force and distance travelled are always perpendicular, no work is done (N.B. remember work = force × distance in the direction of the force). Therefore, the body does not need any energy to be transferred to stay in orbit.

Advanced maths tells us that the larger the orbit radius the slower the body must move because of the weaker gravitational attraction. In addition, the orbit circumference is bigger so the orbital period rapidly increases.

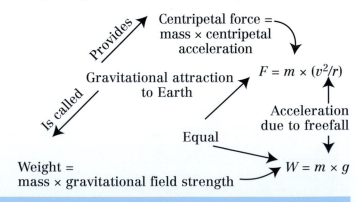

Provides

Centripetal force = mass × centripetal acceleration

$F = m \times (v^2/r)$

Is called

Gravitational attraction to Earth

Acceleration due to freefall

Equal

Weight = mass × gravitational field strength

$W = m \times g$

Questions
1. A planet orbits the Sun. What would happen to the size of its gravitational attraction to the Sun if:
   a. It doubled in mass but stayed in the same orbit?
   b. It stayed the same mass but moved to an orbit twice the distance from the Sun?
2. What happens to the orbital period of a planet as you move away from the Sun? Does the table on p114 confirm this? Give two reasons why the orbital period varies in this way.
3. A geostationary satellite has a mass of 5 kg and an orbit radius of 42×10^6 m.
   a. Show that its orbit circumference is about 260×10^6 m.
   b. Given that its orbital period is 86 400 s, show that its orbital speed is about 3000 m/s.
   c. Therefore, show that the centripetal acceleration is about 0.2 m/s².
   d. Explain why the satellite's weight in this orbit is about 1 N.

OUR PLACE IN THE UNIVERSE The Structure of the Universe

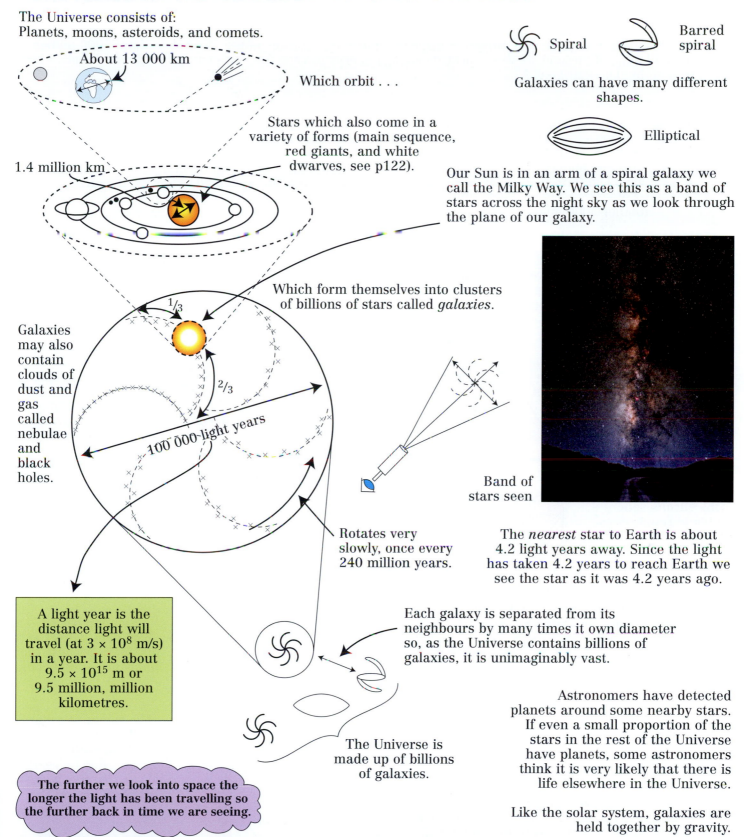

The Universe consists of:
Planets, moons, asteroids, and comets.

About 13 000 km

Which orbit . . .

1.4 million km

Stars which also come in a variety of forms (main sequence, red giants, and white dwarves, see p122).

Galaxies may also contain clouds of dust and gas called nebulae and black holes.

1/3

2/3

100 000 light years

Which form themselves into clusters of billions of stars called *galaxies*.

Spiral

Barred spiral

Galaxies can have many different shapes.

Elliptical

Our Sun is in an arm of a spiral galaxy we call the Milky Way. We see this as a band of stars across the night sky as we look through the plane of our galaxy.

Band of stars seen

Rotates very slowly, once every 240 million years.

A light year is the distance light will travel (at 3×10^8 m/s) in a year. It is about 9.5×10^{15} m or 9.5 million, million kilometres.

The further we look into space the longer the light has been travelling so the further back in time we are seeing.

The Universe is made up of billions of galaxies.

Each galaxy is separated from its neighbours by many times it own diameter so, as the Universe contains billions of galaxies, it is unimaginably vast.

The *nearest* star to Earth is about 4.2 light years away. Since the light has taken 4.2 years to reach Earth we see the star as it was 4.2 years ago.

Astronomers have detected planets around some nearby stars. If even a small proportion of the stars in the rest of the Universe have planets, some astronomers think it is very likely that there is life elsewhere in the Universe.

Like the solar system, galaxies are held together by gravity.

Some astronomers are looking for signals sent by intelligent life from elsewhere in the Universe. This is called the 'search for extraterrestrial intelligence' or SETI.

Questions
1. List the following objects in order of size: galaxy, planet, star, and comet.
2. What force is responsible for holding galaxies together?
3. A galaxy is 100 000 light years from Earth. When we look at the galaxy through a telescope, we are seeing it as it was 100 000 years ago. Explain why.
4. If the nearest star is 4 light years away, show it would take a rocket travelling at 11 km/s (the speed needed to just escape the Earth) about 109 000 years to get there. (Speed of light = 3×10^8 m/s.)
5. Suggest why astronomers find it so difficult to detect planets around stars other than the Sun.

OUR PLACE IN THE UNIVERSE The Sun

For many years, scientists could not work out the source of energy for the Sun. Some thought the energy was released as the Sun shrank in size releasing gravitational potential energy. Others thought it was a chemical reaction like coal burning in a fire. However, geologists knew that the age of the Earth was about 5000 million years old and none of these ideas would provide enough energy to keep the Sun's energy output at the observed rate for anything like that long.

We now know that the Sun is about 4600 million years old and its energy comes from nuclear fusion (see p78 for more details). In the core, under extreme pressure and temperature, hydrogen nuclei are forced together to form helium nuclei releasing vast amounts of energy. There is enough fuel for another 5000 million years.

Einstein's famous relation $\Delta E = \Delta mc^2$, shows this enormous energy release, ΔE, comes at the expense of a small overall loss in the mass of the particles, Δm, linked by the speed of light $c = 3 \times 10^8$ m/s. Inside the Sun, 600 million tonnes of hydrogen are converted in nuclear fusion reactions every second, and 4 million tonnes of this is converted into energy.

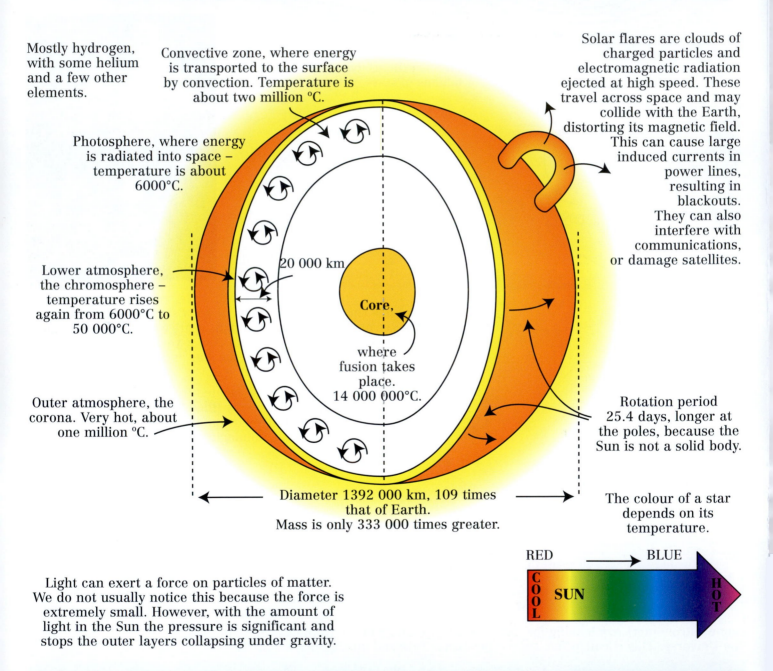

Mostly hydrogen, with some helium and a few other elements.

Convective zone, where energy is transported to the surface by convection. Temperature is about two million °C.

Photosphere, where energy is radiated into space – temperature is about 6000°C.

Lower atmosphere, the chromosphere – temperature rises again from 6000°C to 50 000°C.

Outer atmosphere, the corona. Very hot, about one million °C.

20 000 km

Core, where fusion takes place. 14 000 000°C.

Solar flares are clouds of charged particles and electromagnetic radiation ejected at high speed. These travel across space and may collide with the Earth, distorting its magnetic field. This can cause large induced currents in power lines, resulting in blackouts. They can also interfere with communications, or damage satellites.

Rotation period 25.4 days, longer at the poles, because the Sun is not a solid body.

Diameter 1392 000 km, 109 times that of Earth. Mass is only 333 000 times greater.

The colour of a star depends on its temperature.

Light can exert a force on particles of matter. We do not usually notice this because the force is extremely small. However, with the amount of light in the Sun the pressure is significant and stops the outer layers collapsing under gravity.

RED → BLUE

COOL SUN HOT

Questions
1. What provides the energy for the Sun?
2. How can you guess the temperature of a star simply by looking at it? (N.B. Never look directly at the Sun.)
3. What problems can solar flares cause on Earth?
4. If the Sun looses 4 million tonnes (4×10^9 kg) every second, use $\Delta E = \Delta m \times (3 \times 10^8$ m/s$)^2$ to calculate the energy output of the Sun per second, i.e. its power.
5. Show that if the Sun has a diameter 109× that of the Earth, its volume is over 1 million times greater.

OUR PLACE IN THE UNIVERSE Stars and Their Spectra

Analyzing the light from stars can tell us . . .

- How bright they are (and therefore how far away they are).
- What elements they contain.
- How fast they are moving.

By splitting the starlight up into its spectrum.

(See p43)

Continuous spectrum

Each colour is a unique wavelength.

Starlight from telescope

1. More advanced physics tells us the electrons in Rutherford's model of the atom should lose energy and spiral into the nucleus.

e⁻

2. Clearly, this does not happen, atoms are stable. Niels Bohr suggested that the electrons could only have certain energy levels, like the rungs on a ladder.

3. If an electron is given enough energy, it escapes completely. The atom has been ionized.

4. To move up levels electrons can absorb light energy. Only a precise wavelength, related to the energy difference between levels will be absorbed.

5. Moving down levels releases light of a specific wavelength that depends on the energy difference between levels. Minimum energy; electrons cannot go lower and so cannot crash into the nucleus.

← Energy level; electrons can only have certain energies in the atom, not any energy in between levels, like a person standing on a ladder can only stand on the rungs, not in between.

All hot bodies, like stars, produce a continuous spectrum of wavelengths.

Intensity (brightness)

Hot

Medium

Cool

Wavelength

Every element has a unique set of energy levels and absorbs a unique set of wavelengths.

The wavelength (colour) of the most intense light can be related to temperature.
Cooler stars → redder
Hotter stars → bluer

Temperature is also related to *brightness*.

Cool outer layer

Hot core

Star

Spectrum

Atoms in the outer layer of the star absorb light of certain wavelengths according to the energy levels in the atoms present.

Specific wavelengths not present = black.

Brightness can be used to find the distance to the star (by the inverse square law [see p31]).

This is an absorption line spectrum as certain wavelengths have been absorbed by the outer layers of the star.

Absorption spectra can be thought of as like a fingerprint, or barcode, uniquely identifying the element responsible.

The missing colours are unique to the atoms that absorbed them. Each element produces its own unique *line spectrum*.

The line spectrum identifies what elements are present in the outer layers of the star.

Questions

1. What piece of apparatus could be used to split starlight into its spectrum?
2. Explain how scientists can estimate the temperature of a star by analyzing the light received on Earth from it.
3. Why are there black lines present in the spectra of stars? What information does the position of the lines give us?
4. Here are the line spectra of two stars together with the line spectra of some common elements measured in a laboratory. Which elements are present in each star?

a.

Calcium

Hydrogen

b.

Iron

Magnesium

OUR PLACE IN THE UNIVERSE The Life Story of a Star

A cloud of gas in space is called a *nebula*.

It is made mainly of hydrogen gas, but there may be small amounts of other elements and dust.

1. The gas tends to be quite cool, preventing it dispersing into space.

2. Sometimes the density in parts of the nebula increases.

Billions of years

3. Other molecules are gravitationally attracted to this denser region.

4. The molecules moving towards this region gain kinetic energy (like a ball falling to Earth). Therefore, as the nebula contracts the gas molecules gain kinetic energy. Kinetic energy of a collection of gas molecules is usually referred to as heat energy.

Another way of thinking about this is to consider that the gravitational attraction is compressing the gas, and from the gas laws (p66) the temperature rises proportionally to the pressure.

Fusion
Nuclei are moving so fast that when they collide their electrostatic repulsion cannot keep them apart and they join.

Vast amounts of energy released.

5. The nebula becomes a *protostar*.

Temperature rise to 10

6. The nebula begins to heat up.

Black holes sometimes attract matter from a neighbouring star. This spirals into the black hole, like water down a plughole. This releases X-rays that astronomers can use to detect their presence.

The debris can form a nebula, from which new stars, like our Sun, may form.

X-rays

Blackhole

This is a region of space where, because a very large mass has been compressed into a tiny (possibly zero) volume, gravity is so strong not even light can travel fast enough to escape from it.

This massive explosion releases so much energy that the outer layers are blown off and their nuclei are ripped apart and reformed in every possible way, making the whole range of elements in the periodic table.

Eventually can be implodes

Supernova

Briefly as bright as a whole galaxy.

If the mass of the neutron star is large enough even it will collapse under gravity to form a black hole.

The remaining core is formed entirely of neutrons. It is a *neutron star*.

Neutron star

The matter here is incredibly dense; 1 cm^3 has a mass of about 10 000 000 tonnes.

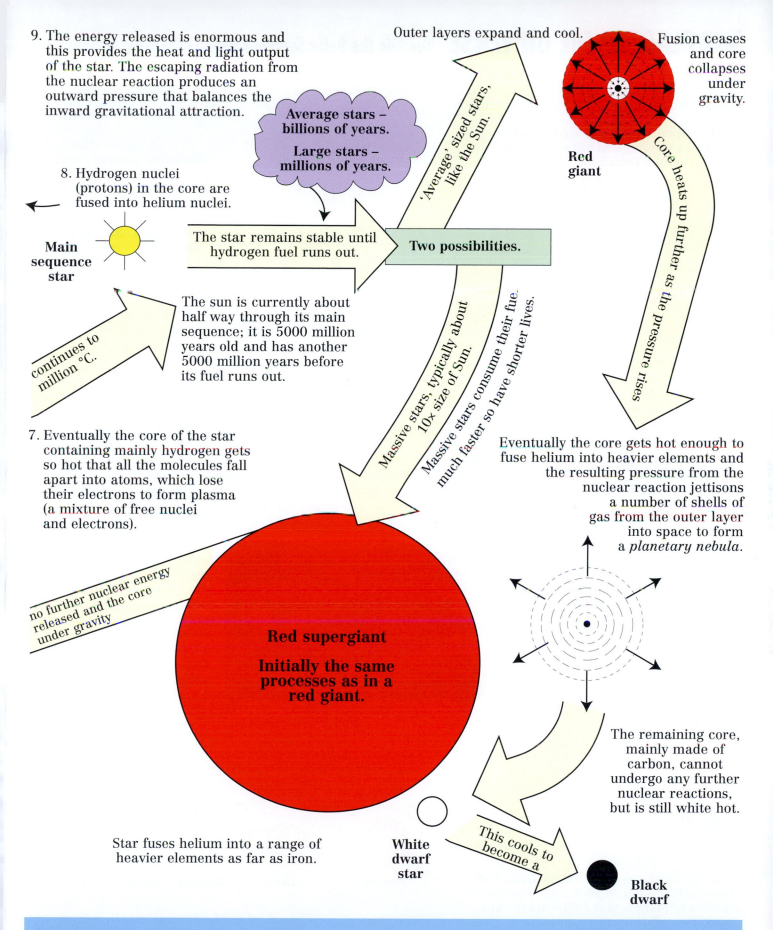

9. The energy released is enormous and this provides the heat and light output of the star. The escaping radiation from the nuclear reaction produces an outward pressure that balances the inward gravitational attraction.

Outer layers expand and cool.

Fusion ceases and core collapses under gravity.

Average stars – billions of years.

Large stars – millions of years.

8. Hydrogen nuclei (protons) in the core are fused into helium nuclei.

Main sequence star

The star remains stable until hydrogen fuel runs out.

Two possibilities.

'Average' sized stars, like the Sun.

Red giant

Core heats up further as the pressure rises

continues to million °C.

The sun is currently about half way through its main sequence; it is 5000 million years old and has another 5000 million years before its fuel runs out.

Massive stars, typically about 10x size of Sun.

Massive stars consume their fuel much faster so have shorter lives.

7. Eventually the core of the star containing mainly hydrogen gets so hot that all the molecules fall apart into atoms, which lose their electrons to form plasma (a mixture of free nuclei and electrons).

Eventually the core gets hot enough to fuse helium into heavier elements and the resulting pressure from the nuclear reaction jettisons a number of shells of gas from the outer layer into space to form a *planetary nebula*.

no further nuclear energy released and the core under gravity

Red supergiant

Initially the same processes as in a red giant.

The remaining core, mainly made of carbon, cannot undergo any further nuclear reactions, but is still white hot.

Star fuses helium into a range of heavier elements as far as iron.

White dwarf star

This cools to become a

Black dwarf

Questions

1. What is an interstellar gas cloud called?
2. Outline the history of the Sun from its formation to its current state.
3. What process provides the energy to make stars shine and stop them collapsing under gravity?
4. Outline what will happen to the Sun when it runs out of hydrogen fuel in its core.
5. What type of star will end in supernova and what might happen to the debris from this explosion?
6. What is a black hole? Can we see them?
7. The early universe only contained hydrogen. Where did all the other elements we see around us come from?

OUR PLACE IN THE UNIVERSE   How Did the Solar System Form?

We saw on p122 that the Sun began to form when a nebula (of gas and dust) collapsed under gravity. The centre of the nebula began to heat up until about 4500 million years ago, when the temperature was high enough, fusion started, and the Sun became a star.

1. Orbiting around the newly formed Sun was the remains of the gas and dust from the nebula.

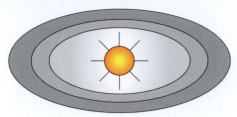

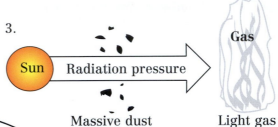

There is evidence for radiation pressure because comets tails, formed from gas like the gas in the early solar system, always point away from the Sun as the gas is blown away by the radiation from the Sun.

2. The debris' gravitational attraction to the Sun kept it in orbit, but the pressure from the radiation escaping from the Sun pushed the lighter gases into a larger orbit, leaving more massive dust particles orbiting closer to the Sun.

3.

Massive dust particles

Light gas particles

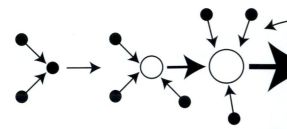

And so on

The heat generated by these collisions melted the rocks allowing the young planets to form into spheres before they cooled down.

4. The dust particles collided with each other and began to collect into larger clumps. These grew as they collected more dust into rocks, which eventually joined to form planets.

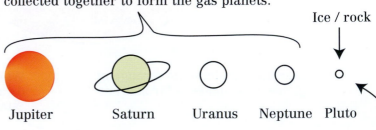

Craters on other planets are evidence for these collisions, which still continue. We call the small rocks asteroids. Plate tectonics covers up craters on Earth, but there are still some impact craters to be seen. Scientists think these may explain some extinction processes such as that of the dinosaurs.

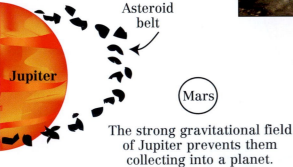

Asteroid belt

Jupiter

Mars

The strong gravitational field of Jupiter prevents them collecting into a planet.

Some astronomers think the asteroid belt, between Mars and Jupiter, might be the remains of planets that collided, perhaps due to the influence of Jupiter's very strong gravitational field.

The gases further out in the solar system also collected together to form the gas planets.

Ice / rock

Jupiter   Saturn   Uranus   Neptune   Pluto

Pluto does not fit this pattern. It is suggested that it has been captured by the Sun's gravity and did not form in the solar system.

OUR PLACE IN THE UNIVERSE The Expanding Universe

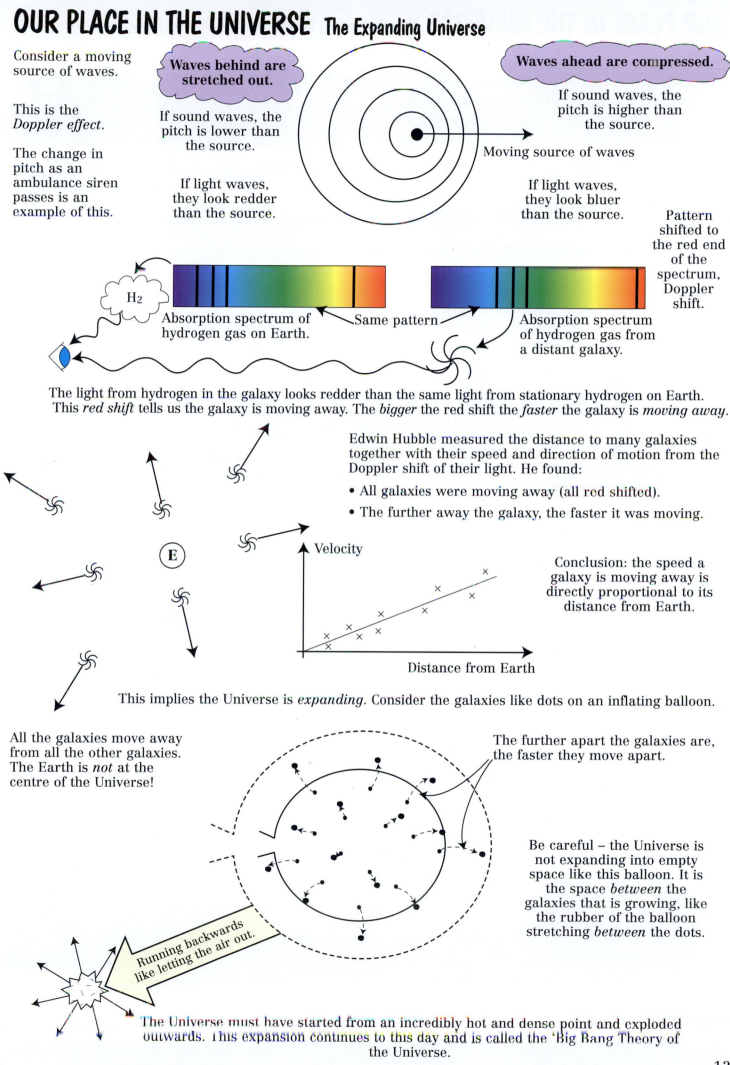

Consider a moving source of waves.

This is the *Doppler effect*.

The change in pitch as an ambulance siren passes is an example of this.

Waves behind are stretched out.

If sound waves, the pitch is lower than the source.

If light waves, they look redder than the source.

Waves ahead are compressed.

If sound waves, the pitch is higher than the source.

Moving source of waves

If light waves, they look bluer than the source.

Pattern shifted to the red end of the spectrum, Doppler shift.

H₂

Absorption spectrum of hydrogen gas on Earth.

Same pattern

Absorption spectrum of hydrogen gas from a distant galaxy.

The light from hydrogen in the galaxy looks redder than the same light from stationary hydrogen on Earth. This *red shift* tells us the galaxy is moving away. The *bigger* the red shift the *faster* the galaxy is *moving away*.

Edwin Hubble measured the distance to many galaxies together with their speed and direction of motion from the Doppler shift of their light. He found:

- All galaxies were moving away (all red shifted).
- The further away the galaxy, the faster it was moving.

E

Velocity

Distance from Earth

Conclusion: the speed a galaxy is moving away is directly proportional to its distance from Earth.

This implies the Universe is *expanding*. Consider the galaxies like dots on an inflating balloon.

All the galaxies move away from all the other galaxies. The Earth is *not* at the centre of the Universe!

The further apart the galaxies are, the faster they move apart.

Be careful – the Universe is not expanding into empty space like this balloon. It is the space *between* the galaxies that is growing, like the rubber of the balloon stretching *between* the dots.

Running backwards like letting the air out.

The Universe must have started from an incredibly hot and dense point and exploded outwards. This expansion continues to this day and is called the 'Big Bang Theory of the Universe.

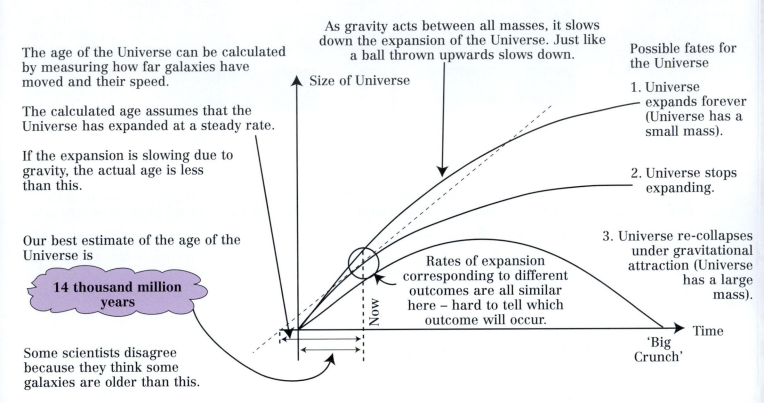

Space is full of microwaves.

Big Bang
Hot and dense
Lots of high energy light

Expansion and cooling

Now
Light from Big Bang has lost energy and been stretched to microwaves.

About 10% of the 'snow' on an untuned TV.

This Cosmic microwave background is good evidence for the Big Bang Theory.

The age of the Universe can be calculated by measuring how far galaxies have moved and their speed.

The calculated age assumes that the Universe has expanded at a steady rate.

If the expansion is slowing due to gravity, the actual age is less than this.

Our best estimate of the age of the Universe is

14 thousand million years

Some scientists disagree because they think some galaxies are older than this.

As gravity acts between all masses, it slows down the expansion of the Universe. Just like a ball thrown upwards slows down.

Size of Universe

Now

Rates of expansion corresponding to different outcomes are all similar here – hard to tell which outcome will occur.

Possible fates for the Universe

1. Universe expands forever (Universe has a small mass).

2. Universe stops expanding.

3. Universe re-collapses under gravitational attraction (Universe has a large mass).

Time

'Big Crunch'

Other theories

Which of fates 1, 2, or 3 will occur depends on the total mass of the Universe and the rate of expansion.

We need to calculate the mass of the Universe.

But we think the majority of the mass of the Universe is 'dark matter', which we cannot see.

Measuring this and the exact rate of expansion is very hard so the age and fate of the Universe remain very controversial.

Oscillating Universe

Steady State

Universe expands

New matter generated in the space created.

Size of Universe

Big Crunch followed by another Big Bang.

Universe continually expanding and contracting.

Time

Questions

1. What is the Doppler effect? Suggest where you might be able to observe the Doppler effect in everyday life.
2. If a galaxy was moving towards the Earth, how would the light received from it be affected? What if it was moving away?
3. Explain what led Hubble to propose that the Universe is expanding.
4. What was the Big Bang? Suggest two pieces of evidence for this theory of the Universe.
5. Outline three possible fates for the Universe. What two factors will dictate which outcome actually occurs?
6. Suggest some reasons why scientists are uncertain about the age and the fate of the Universe.
7. Make a list of three controversial facts in this topic. Explain why they are controversial. If possible, suggest some data scientists could collect to try to settle the dispute.

FORMULAE
The lists below bring together all the formulae in the book.

FORCES AND MOTION
Speed (m/s) = distance (m) / time (s) $s = d/t$

Average speed (m/s) = total distance travelled (m) / total time taken (s) $s = d/t$

Acceleration (m/s^2) = change in velocity (m/s) / time taken (s) $a = \Delta v/\Delta t$

Equations of motion for uniformly accelerated motion

$$v = u + at$$
$$x = ut + \tfrac{1}{2}at^2$$
$$v^2 = u^2 + 2ax$$

v = final velocity (m/s)
u = initial velocity (m/s)
a = acceleration (m/s^2)
t = time taken (s)
x = distance travelled (m)

Force (N) = mass (kg) × acceleration (m/s^2). $F = ma$

Weight (N) = mass (kg) × acceleration due to gravity (m/s^2). $W = mg$

Weight (N) = mass (kg) × gravitational field strength (N/kg). $W = mg$

Density (kg/m^3) = mass (kg) / volume (m^3) $D = m/V$

Pressure (N/m^2 or Pa) = force (N) / area (m^2) $P = F/A$

Momentum (kgm/s) = mass (kg) × velocity (m/s)

Impulse (Ns) = Force (N) × time force acts for (s) = change in momentum (kgm/s) $F\Delta t = mv - mu$

Centripetal acceleration (m/s^2) = [velocity (m/s)]^2 / radius (m). $a = v^2/r$

Centripetal force = mass (kg) × acceleration (m/s^2) = mass (kg) × [velocity (m/s)]^2 / radius (m) $F = mv^2/r$

Orbital speed (m/s) = orbit circumference (m) / orbital period (s) $v = 2\pi r/T$

Moment (Nm) = Force (N) × perpendicular distance from line of action of the force to the axis of rotation (m).

Principle of moments
Sum of anticlockwise moments = sum of clockwise moments when in equilibrium.

Energy
Work done = force (N) × distance moved in the direction of the force (m). $w.d. = F \times d$

Power (W) = energy transferred (J) / time taken (s). $P = E/t$

Energy transferred = work done

Gravitational potential energy transferred (J) = mass (kg) × gravitational field strength (N/kg) × change in height (m) $GPE = mg\Delta h$

Kinetic energy (J) = $1/2$ mass of object (kg) × [speed (m/s)]^2. $KE = \times mv^2$

Efficiency (%) = useful energy output (J) / total energy input (J) × 100%.

Nuclear energy
Energy released (J) = change in mass (kg) × [speed of light (m/s)^2] $\Delta E = \Delta mc^2$

Waves
Wave speed (m/s) = frequency (Hz) × wavelength (m). $v = f\lambda$

Intensity (W/m^2) = power (W) / area (m^2). $I = P/A$

Refractive index, n = speed of light in vacuum (m/s) / speed of light in medium (m/s) $n = c/v$

Snell's Law

Refractive index n, = sin (angle of incidence) / sin (angle of refraction) $n = \sin i / \sin r$

sin (critical angle) = refractive index of second material / refractive index of first material $\sin c - n_r / n_i$

Magnification = image height / object height

Power of lens (dioptre) = 1/focal length (metres)

Angular magnification = focal length of objective lens / focal length of eyepiece lens

Electricity
Current (A) = charge passing (C) / time taken (s). $I = Q/t$

Potential difference (V) = energy transferred (J) / charge passing (C). $V = E/Q$

Resistance (Ω) = potential difference (V) / current (A) $R = V/I$

Power (W) = [current (A)]^2 × resistance (Ω) $P = I^2R$

Power (W) = current (A) × voltage (V) $P = IV$

Power (W) = [voltage (V)]^2 / resistance (Ω) $P = V^2/R$

Electrical energy (kWh) = power (kW) × time (h)

Kinetic energy of an electron (J) = charge on the electron (C) × potential difference (V) $KE = e \times V$

Transformer formula

Primary voltage (V) / secondary voltage (V) = No. of turns on primary / No. of turns on secondary. $V_p/V_s = N_p/N_s$

Thermal physics
Kelvin → °C = (temperature / K) − 273

°C → Kelvin = (temperature / °C) + 273

Energy supplied (J) = mass (kg) × specific heat capacity (J/kg K) × temperature change (K) $\Delta E = m \times s.h.c. \times \Delta T$

Energy (J) = mass (kg) × specific latent heat (J/kg) $E = mL$

Pressure (Pa) / temperature (Kelvin) = constant. $P/T = $ constant.

Pressure (Pa) × volume (m^3) = constant. $PV = $ constant

Units
Length – metres, m
Time – seconds, s
Mass – kilogram, kg
Speed or velocity – metres per second, m/s
Acceleration – metres per second^2, m/s^2
Force – Newton, N
Momentum – kilogram metre per second, kgm/s
Impulse – Newton second, Ns
Moment – Newton metre, Nm
Density – kilograms per metre^3, kg/m^3
Pressure – Newton per metre^2, N/m^2 (equivalent to 1 Pascal, Pa)
Work done – Newton metre, Nm
Power – Watt, W
Energy – Joule, J (equivalent to one Newton metre, Nm)
Frequency – Hertz, Hz
Wavelength – metre, m
Intensity – Watts per metre^2, W/m^2
Power of lens – dioptre
Current – Amps, A
Charge – Coulombs, C
Potential difference – Volts, V
Resistance - Ohms, Ω
Electrical energy - Joules, J (or kiloWatt-hours, kWh. 1 kWh = 3.6×10^6 J)
Temperature – Kelvin, K or Celsius, °C.
Specific heat capacity – Joules per kilogram per Kelvin, J/kg K
Specific latent heat – Joules per kilogram, J/kg

INDEX